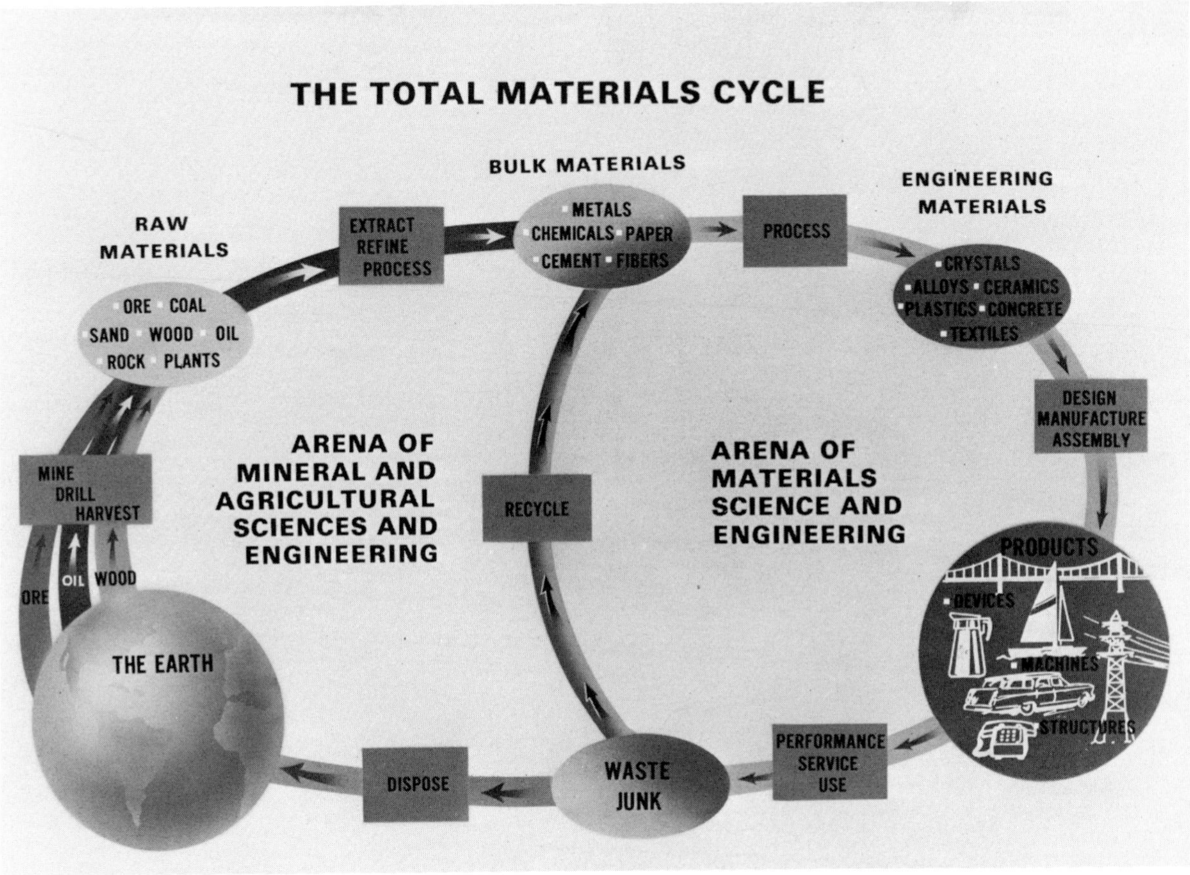
THE TOTAL MATERIALS CYCLE
BULK MATERIALS
RAW MATERIALS
ENGINEERING MATERIALS
EXTRACT REFINE PROCESS
METALS CHEMICALS PAPER CEMENT FIBERS
PROCESS
ORE COAL SAND WOOD OIL ROCK PLANTS
CRYSTALS ALLOYS CERAMICS PLASTICS CONCRETE TEXTILES
DESIGN MANUFACTURE ASSEMBLY
MINE DRILL HARVEST
ARENA OF MINERAL AND AGRICULTURAL SCIENCES AND ENGINEERING
RECYCLE
ARENA OF MATERIALS SCIENCE AND ENGINEERING
PRODUCTS
DEVICES
MACHINES
STRUCTURES
ORE
OIL
WOOD
THE EARTH
PERFORMANCE SERVICE USE
WASTE JUNK
DISPOSE

Materials and Man's Needs

Materials Science and Engineering

Summary Report of the
Committee on the Survey of Materials Science
and Engineering

NATIONAL ACADEMY OF SCIENCES
Washington, D.C.
1974

NOTICE

This report is one of a series of survey reports on major areas of science and technology, sponsored and organized by the Committee on Science and Public Policy of the National Academy of Sciences.

The members of the committee selected to undertake this project and prepare this report were chosen for recognized scholarly competence and with due consideration for the balance of the disciplines appropriate to the project. Responsibility for the detailed aspects of this report rests with the committee.

Each report issuing from the National Academy of Sciences is reviewed by an independent group of qualified individuals according to procedures established and monitored by the Report Review Committee of the National Academy of Sciences. Distribution of the report is approved, by the President of the Academy, upon satisfactory completion of the review process.

Frontispiece: A schematic illustration of the materials cycle, showing the principal arena of materials science and engineering. Not indicated are myriad possible subcycles, including those related to the environment and energy.

ISBN 0-309-02220-7

Library of Congress Catalog Card Number 74-2118

Printed in the United States of America

December 1973

Dear Dr. Handler:

The Committee on Science and Public Policy takes pleasure in transmitting to you the Summary Report of its Committee on the Survey of Materials Science and Engineering (COSMAT) under the title Materials and Man's Needs. This committee was originally appointed by you in December of 1970 under the chairmanship of Dr. Morris Cohen of The Massachusetts Institute of Technology.

The present report may be considered as the latest in the series of reports on the needs and opportunities of the various scientific disciplines sponsored by the Committee on Science and Public Policy (COSPUP). However, it differs from earlier reports in the series in several important respects. Materials science and engineering is not a single discipline, like physics or chemistry, but rather a multi-disciplinary area of research and engineering that has evolved gradually over the last twenty years into a coherent aggregation of related activities, but within which each of the component disciplines retains its identity. Although the earlier disciplinary reports laid considerable stress on the applications of each discipline to national needs in general, the principal emphasis was on the scientific opportunities in the discipline and on the means required to realize and exploit these opportunities.

The present report lays more stress on the needs themselves, and on their implications for the organization and evolution of research and education in the field of materials. There is particular emphasis on the engineering and design aspects of the field. This difference in emphasis is reflected in the composition of the Committee; out of 23 members, 13 come from non-academic institutions, and several, including one economist, one historian of technology, and one geologist, are not specialists in materials research.

The report attempts to advance further the effort made in earlier COSPUP reports to find ways of identifying priorities in a field of research. A rather elaborate survey was designed in which nearly 1,000 specialists from a wide variety of technical fields participated. The purpose of this survey was to relate priorities within various technical subfields to various areas of future national needs, such as energy, transportation, communications, environmental protection, and public health. As often seems to be true in such studies, some of the results are ambiguous, and their translation into governmental or other institutional policies are not obvious. Nevertheless, a few patterns emerge, and we feel that the exercise will significantly advance public discussion of priorities in general as well as specific priorities for materials research.

The appearance of the COSMAT report seems especially timely in relation to other national reports in the materials field. The last few years has witnessed rising public and Congressional concern with the future availability of natural resources for the economic growth of the United States, and with the development of more coherent national strategies for the wise and efficient use of available resources. With this concern there has emerged the concept of the "materials cycle," the life history of materials from extraction of the original raw materials through design of a product to final disposal of the used product into the environment, or to recycling for other uses. This concept constitutes the focus and theme of the present report. The report should be read in conjunction with the recently published report of the National Commission on Materials Policy and with the reports of the Secretary of the Interior mandated by the Mining and Minerals Policy Act of 1970, as well as with several privately sponsored reports, such as the recent report of the American Chemical Society, Chemistry in the Economy, and the National Academy of Sciences' earlier report, Man, Materials, and the Environment.

Although placing greater emphasis on the translation of national needs into scientific priorities, the present report does not overlook the opportunity-oriented aspects of the materials field. The chapter entitled "Opportunities in Materials Research" provides an exciting account of new developments in the field, and points out the many opportunities for new conceptual advances as well as new progress in materials processing and in development and study of whole new classes of materials.

Very truly yours,

Melvin Calvin
Chairman
Committee on Science and
Public Policy

This report, <u>Materials and Man's Needs</u>, is the latest in a
series of survey reports concerning major fields of science and
technology, prepared under the aegis of the Academy's Committee
on Science and Public Policy. In this instance, the report presents
an intensive examination of a field that is inherently multidisciplinary
and that impinges in innumerable ways upon economic stability and the
quality of life.

Materials science and technology is unquestionably a resource
of the greatest importance. The Academy is indebted to the Committee
on the Survey of Materials Science and Engineering for this substantial
contribution to our appreciation and understanding of that resource.

Philip Handler
President

Washington, D.C.
December 1973

PREFACE

What is materials science and engineering? How does it function? How can it contribute more effectively to the achievement of societal goals and national purpose? This report, Materials and Man's Needs: Materials Science and Engineering, attempts to answer these and related questions.

In December 1970, the President of the National Academy of Sciences appointed a Committee on the Survey of Materials Science and Engineering (COSMAT) to conduct a comprehensive analysis and assessment of the field of materials science and engineering. This Survey follows others in various fields of science conducted during the past decade under the aegis of the Academy's Committee on Science and Public Policy. However, the COSMAT study is the first to embrace both science and engineering in its coverage, centering on the relationships between the structure and properties of materials as a basis for their design, preparation, and utilization by mankind.

The present report summarizes the main results of the COSMAT study and is intended primarily for executives and administrators in government, industry, and universities who are involved with decisions or policies in which materials and the associated sciences and technologies play a significant role. At the same time, some parts of the report are designed to portray the intellectual excitement of the field and will be of special interest to the working scientist or engineer. The relatively concise form of this Summary Report will facilitate access to the principal findings and recommendations of the Survey, as well as to COSMAT's priority analysis of materials research bearing on national goals and programs. A more complete account of the COSMAT study will appear subsequently.

The overall objectives of the Survey have been: to determine the nature and scope of materials science and engineering; to ascertain the linkage of science with engineering in the field of materials for the successful translation of new basic knowledge into useful application; to examine the interaction of materials science and engineering with other areas of science and technology; to discern trends in the development of the materials field in order to identify its challenges, opportunities, and needs; and to reach conclusions concerning the means by which materials science and engineering might contribute more broadly to the national well-being. The Survey Committee was constituted deliberately to reflect not only the wide range of scientific and engineering disciplines comprising the field itself, but also the experience of practitioners in universities, industry, and government. Moreover, in order to obtain a somewhat "outside view" as well, professionals in economics, history, geology, and engineering design were included in the Committee membership.

Because materials occupy a central position in national economies as well as in man's daily life, COSMAT found it appropriate, indeed essential, to direct considerable attention to the larger field of materials in which materials science and engineering operate. Similarly, research priorities in materials science and engineering were addressed not only from the standpoint of creative appeal, but also in the context of societal goals, ecological constraints, and dwindling natural resources. To obtain necessary information for this broad approach, COSMAT appointed 11 panels and committees and secured responses through questionnaires and position papers from a wide community of experts. Nearly 1,000 persons and institutions provided inputs on various aspects of the Survey and more than 100 individuals participated directly in the COSMAT program.

The field of materials science and engineering has turned out to be so extensive in scope and so pervasive in its applications that we cannot claim to have elucidated all of its important manifestations sufficiently. But the COSMAT Survey does provide a framework for delineating those areas that may merit further examination. Partly by design and partly from inability at this stage, COSMAT has not attempted to make specific estimates of the funding and manpower needed for fulfilling the role of materials science and engineering in the nation. That further inquiry will come about naturally due to societal driving forces, we are convinced, if COSMAT has done justice to its exploratory mission. Related studies that help round out the panorama of the materials field are the Academy reports on "Minerals Science and Technology: Needs, Challenges and Opportunities" (1969); "Physics in Perspective" (1972); "Man, Materials, and the Environment" (1973); the report of the American Chemical Society on "Chemistry in the Economy" (1973); and the final report of the National Commission on Materials Policy (1973). The COSMAT report, in particular, points out the role of materials science and engineering within the framework of that policy.

COSMAT is appreciative of the foresight of the Committee on Science and Public Policy in discerning the timeliness and importance of surveying the field of materials science and engineering, with its interrelations among technology, science, and society. Harvey Brooks, then Chairman of the Committee on Science and Public Policy, and Robert E. Green, its Executive Secretary, were most helpful in getting this study launched under the auspices of the National Academy of Sciences.

COSMAT is also particularly grateful to the National Science Foundation and the Advanced Research Projects Agency for their financial and cooperative support of the COSMAT undertaking. The valuable assistance received from many individuals and professional organizations in the United States and abroad added immeasurably to the depth and comprehensiveness of the study. We also acknowledge the very perceptive and skillful efforts of its consultant, Kenneth M. Reese, in the writing of this report.

Morris Cohen, Chairman

William O. Baker, Vice Chairman

Committee on the Survey of
Materials Science and Engineering

COMMITTEE ON THE SURVEY OF

MATERIALS SCIENCE AND ENGINEERING (COSMAT)

*Morris Cohen (Chairman)	Massachusetts Institute of Technology
*William O. Baker (Vice Chairman)	Bell Telephone Laboratories, Inc.
Donald J. Blickwede	Bethlehem Steel Corporation
Raymond F. Boyer	Dow Chemical Company
*Paul F. Chenea	General Motors Corporation
Preston E. Cloud	University of California, Santa Barbara
*Daniel C. Drucker	University of Illinois
Julius J. Harwood	Ford Motor Company
I. Grant Hedrick	Grumman Aerospace Corporation
Walter R. Hibbard, Jr.	Owens-Corning Fiberglas Corporation
*John D. Hoffman	National Bureau of Standards
Melvin Kranzberg	Georgia Institute of Technology
*Hans H. Landsberg	Resources for the Future, Inc.
Humboldt W. Leverenz	RCA Laboratories, Inc.
Donald J. Lyman	University of Utah
Roger S. Porter	University of Massachusetts
Rustum Roy	Pennsylvania State University
*Roland W. Schmitt	General Electric Company
Abe Silverstein	Republic Steel Corporation
Lawrence H. Van Vlack	The University of Michigan

Ex-Officio Members

*Harvey Brooks   Harvard University
 (as former Chairman, Committee on
 Science and Public Policy, NAS)

*N. Bruce Hannay   Bell Telephone Laboratories, Inc.
 (as Chairman, National Materials
 Advisory Board, National Research
 Council, NAS-NAE)

*Ernst Weber   National Academy of Sciences
 (as Chairman, Division of
 Engineering, National Research
 Council, NAS-NAE)

*Members of the Executive Board

Survey Directors

Alan G. Chynoweth	Bell Telephone Laboratories, Inc.
S. Victor Radcliffe	Case Western Reserve University

COMMITTEE ON SCIENCE AND PUBLIC POLICY

TABLE OF CONTENTS

FIGURES

TABLES

THE MESSAGE OF COSMAT

Materials have been ingrained in human culture since the beginning of history. They rank with energy and information as basic resources of mankind. Both nature-given and man-made materials are the working substances of our civilization, and they should be probed, manipulated, modified, utilized, and safeguarded with due respect. Considerations like these stimulate the intellectual and social forces that are tending to draw a science and engineering of materials into being. But though the style of activity in this field may be new, its existence in one form or another may long have been appreciated. The Greeks had a word that seems apt today; "hylology, a doctrine or science of matter."

Materials science and engineering is emerging, COSMAT believes, as a coherent doctrine or technical field with deep intellectual roots, which promises new contributions, on a practical time scale, to the nation's prosperity, security, and quality of life. It intimately combines knowledge of the condensed state of matter with the real world of material function and performance. It links the quest for deep fundamental understanding of matter with the imperative of satisfying man's needs. Over all, materials science and engineering is a purposeful enterprise closely coupled to mankind's requirements for products, structures, machines, and devices. Herein lies its strength, value, and novelty. In a way perhaps unprecedented in the history of science and technology, it presents a basis for sophisticated management of a total field of science and engineering that leads at the same time to benefits for society and to professional satisfactions for its practitioners.

Society will always need material things but in addition there is growing worldwide concern over the continued supply of energy and the impact of man's technology on the environment: the challenges to materials science and engineering have never been greater. These concerns are compounded by an equally growing awareness of the difficulties in assuring the continued availability of raw materials. It is vital to learn how to move materials carefully around the materials cycle, from production of raw materials to use and eventual disposal, in ways that minimize strain on natural and environmental resources. To help improve the management of all aspects of materials, the concepts and methods of materials science and engineering must be applied systematically at all points in the materials cycle and coordinated with evolving national policies on resources, energy, and environmental quality.

Consistent with this central theme our design has five general thrusts:

- Purposeful mobilization of the technical expertise of materials scientists and engineers.

- Comprehensive federal leadership.

- An intensive effort by industry to exploit and nourish the reservoir of materials knowledge.

- A strengthened academic base to inject new knowledge into that reservoir.

- A deliberate assumption of pertinent responsibilities by materials professionals and their technical societies.

Specific technical goals for materials science and engineering are not difficult to find. Virtually all our current and prospective methods of generating, transmitting, storing, and using energy are materials-limited, some of them seriously so. Many stresses on environmental quality caused by man's technological operations can be alleviated only by imaginative and comprehensive materials technology. Materials concepts should also be applied systematically to favor the use of abundant and renewable resources and to upgrade the recyclability of materials, particularly through materials selection and development, and product design. Technical attention is invited by many other areas, including transportation, biomaterials, advanced automation technologies, and construction materials. Certain materials problems are generic to many of these areas and call for intensified research backed up by longer-range research on fundamental properties of materials rather more complex than those customarily studied. To these ends, we shall be recommending that:

1. Critical materials research and development be given high emphasis in national programs to alleviate energy shortages.

2. The interdisciplinary capabilities of materials science and engineering be systematically brought to bear on problems of environmental quality.

3. Materials research and development be recognized as vital to achieving specific technical goals and be adequately supported.

4. Important opportunities for applied materials research of broad implication be tackled more vigorously.

5. Fundamental research in materials be broadened to address more complex materials.

6. The feasibility of using forests and other renewable organic sources as raw materials for plastics be assessed.

7.   Resources of materials science and engineering be invoked
     to increase the recyclability of materials and products.

The necessity for an overall, creative, and prudent management
of our materials resources toward a closed cycle of efficient use and
reuse is unquestioned today. Every contemporary materials study docu-
ments this need in terms of provision for the future, advancing the
national economy, restoring and securing the national energy balance,
and improving as well as conserving the environment. Federal leadership
is required to achieve these ends. This leadership should be coherent
and oriented toward establishing materials policies in harmony with
those on resources, energy, and the environment. It should draw on the
technical community, as necessary, for the analytical and advisory tools
required, particularly to develop a more quantitative understanding of
the materials cycle. And it should work along these lines with its
counterparts in other nations and in international bodies, since
neither knowledge nor resources are the exclusive concern of the United
States. In the federal infrastructure for materials science and engineering research programs, there needs to be balanced support for basic
and applied research, effective coordination among various agencies
and departments, and optimal utilization of federal research laboratories.
To these purposes, we shall be recommending that:

8.   The federal government develop a broad materials policy
     in the context of the materials cycle and on the same level
     as related policies for energy and the environment.
9.   Urgent attention be given to developing international
     cooperation in the materials field.

10. The allocation of resources between basic and applied
     research in materials be continually examined and kept
     appropriate to national needs.

11. There be effective coordination of materials research
     programs among federal agencies and departments.

12. The opportunities for materials research afforded by
     federal laboratories be used to the full.

In the industrial sector there is impressive potential for
fuller involvement of materials science and engineering. There are
rich opportunities for new products to meet societal demands, in harmony
with new processes to make more efficient use of available resources.
However, such advances will require a renewed commitment to materials
research and development, a reasoned combination of basic and applied
research, and a coordinated application of materials knowledge as
demonstrated successfully by the higher-technology industries. Small
companies could benefit significantly from joint sponsorship of generic
applied research on materials, particularly at universities where the
appropriate talents and facilities happen to exist; there are also
sound arguments for governmental participation in this connection.
Specifically, we will recommend that:

13. Industry work to integrate materials science and
     engineering with product design and manufacture.

14. Cooperation between industries in materials research
     and development be expedited.

15. Mature industries in the materials field strengthen their research, particularly in materials processing and manufacturing.

16. Industry, government, and universities cooperate in establishing needed materials research facilities.

The bodies of knowledge required for progress in materials frequently do not coincide with those of the traditional disciplines. Though the latter contribute significantly to materials science and engineering, they tend to maintain a compartmentalization of the field. It is incumbent on universities to seek a more flexible balance of disciplinary, multidisciplinary and interdisciplinary activities in both education and research, and the field of materials offers fertile ground for such evolutionary changes. We believe that materials science will play an increasingly significant part in the education and work of physical and life scientists as well as of engineers and technologists. Accordingly, we shall recommend that:

17. Universities intensify their efforts to build interdisciplinary activities in research and education.

18. Solid-state topics play a more significant part in the undergraduate education of physical scientists as well as of engineers.

19. Undergraduate curricula in materials be designed to strengthen the role of engineering.

20. Interdisciplinary materials research laboratories continue to receive a substantial proportion of their support as

block funds and continue to evolve techniques for effective local management of these funds.

In the light of these trends and challenges, scientific and engineering societies concerned with materials should vigorously pursue initiatives for the effective coordination of programming, journals, information services, and all such professional matters that will help the field contribute its full potential to human well-being, national purpose, and to science and engineering generally. Lively interplay among the professions is just as indispensable as among the disciplines, and the technical societies should provide a fluid medium for this interplay. To these various ends we shall recommend that:

21. The National Research Council work toward taking full advantage of its opportunites to draw on expertise in the materials community.

22. Professional societies in the materials field deliberately seek to coordinate their activities.

23. Government, industry, and universities develop arrangements for personnel interchanges and interactions and make the fullest possible use of existing ones.

24. Improved data- and statistics-gathering mechanisms useful for the multidisciplinary field of materials science and engineering be developed and supported on a continuing basis.

INTRODUCTION

<u>Definitions and Concepts</u>

Materials science and engineering is a multidisciplinary activity that has emerged in recognizable form only during the past two decades. Practitioners in the field develop and work with materials that are used to make things -- products like machines, devices, and structures. More specifically:

> Materials science and engineering is concerned
> with the generation and application of knowledge
> relating the composition, structure, and processing
> of materials to their properties and uses.

The multidisciplinary nature of materials science and engineering is evident in the educational backgrounds of the half-million scientists and engineers who, in varying degree, are working in the field. Only about 50,000 of them hold materials-designated degrees;[*] the rest are largely chemists, physicists, and nonmaterials-designated engineers. Many of these professionals still identify with their original disciplines rather than with the materials community. They are served by some 35 national societies and often must belong to several to cover their

---

[*] We define a "materials-designated degree" as one containing in its title the name of a material or a material process or the word "material." Examples include metallurgy, ceramics, polymer science or engineering, welding engineering, and materials science or engineering. Thus far, virtually all materials-designated degrees are in matallurgy or ceramics.

professional and technical needs. This situation is changing, if slowly. One recent indication was the formation of the Federation of Materials Societies, in 1972. Of the 17 broadly based societies invited to join, nine had done so by October 1973.

Materials are exceptionally diverse. The scope of materials science and engineering spans metals, ceramics, semiconductors, dielectrics, glasses, polymers, and natural substances like wood, fibers, sand, and stone. For our purposes we exclude certain substances that in other contexts might be called "materials." Typical of these are foods, drugs, water, and fossil fuels. Materials as we define them have come increasingly to be classified by their function as well as by their nature; hence, biomedical materials, electronic materials, structural materials. This blurring of the traditional classifications reflects in part our growing, if still imperfect, ability to custom-make materials for specific functions.

The Nature of Materials

Materials, energy, and the environment are closely inter-related. Materials are basic to manufacturing and service technologies, to national security, and to national and international economies. The housewife has seen her kitchen transformed by progress in materials: vinyl polymers in flooring; stainless steel in sinks; Pyroceram and Teflon in cookware. The ordinary telephone contains in its not-so-ordinary components 42 of the 92 naturally occurring elements. Polyethylene, an outstanding insulator for radar equipment, is but one of the myriad materials vital to national defense. By one of several possible reckonings, production and forming of materials account for some 20 percent of the nation's gross national product, but the number is deceptive; without materials we would have no gross national product.

Man tends to be conscious of products and what he can do with them, but also tends to take the materials in products for granted. Nylon is known far better in stockings than as the polyamide engineering material used to make small parts for automobiles. The transistor is known far better as an electronic device, or as a pocket-size radio, than as the semiconducting material used in the device and its many relatives.

Some materials produce effects out of proportion to their cost or extent of use in a given application. Synthetic fibers, in the form of easy-care clothing, have worked startling changes in the lives of housewives. Certain phosphor crystals, products of years of research on materials that emit light when bombarded by electrons, provide color-television pictures at a cost of less than 0.5 percent of the manufacturing cost of the set.

The properties of specific materials often determine whether a product will work. In manned space flights, ablative materials of modest cost are essential to the performance of the heat shield on atmospheric reentry vehicles. New or sharply improved materials are critical to progress in energy generation and distribution. At the other extreme are home-building materials, whose properties, though important, need not be markedly improved to meet society's goals in housing.

Materials commonly serve a range of technologies and tend to be less proprietary than are the products made of them. Materials, as a result, are likely to offer more fruitful ground for research and development, including cooperative research and development, than are specific products. One example is certain "textured" materials, polycrystalline structures in which the alignment of neighboring crystals is determined by the processing steps employed. The ability thus to control crystal orientation grew out of research by physicists, metallurgists, and even mathematicians. The resulting improvements in properties are proving useful in a widening spectrum of applications. They include soft magnetic alloys for memory devices, oriented steels for transformers, high-elasticity phosphor bronze for electrical connectors, and steel sheet for automobile fenders, appliance housings, and other parts formed by deep drawing.

The Total Materials Cycle

All materials move in a "total materials cycle" (Frontispiece), which in this report we will simply call the "materials cycle." From the earth and its atmosphere man takes ores, hydrocarbons, wood, oxygen, and other substances in crude form and extracts, refines, purifies, and converts them into simple metals, chemicals, and other basic raw materials. He modifies these raw materials to alloys, ceramics, electronic materials, polymers, composites, and other compositions to meet performance requirements; from the modified materials he makes shapes or parts for assembly into products. The product, when its useful life is ended, returns to the earth or the atmosphere as waste. Or it may be dismantled to recover basic materials that reenter the cycle.

The materials cycle is a global system whose operation includes strong three-way interactions among materials, the environment, and energy supply and demand. The condition of the environment depends in large degree on how carefully man moves materials through the cycle, at each stage of which impacts occur. Materials traversing the cycle may represent an investment of energy in the sense that the energy expended to extract a metal from ore, for example, need not be expended again if the metal is recycled. Thus a pound of usable iron can be recovered from scrap at about 20 percent of the "energy cost" of extracting a pound of iron from ore. For copper the figure is about 5 percent, for magnesium about 1.5 percent.

Materials scientists and engineers work most commonly in that part of the materials cycle that extends from raw materials through dismantling and recycling of basic materials. Events in this (or any other) area typically will have repercussions elsewhere in the cycle or system. Research and development, therefore, can open new and sometimes surprising paths around the cycle with concomitant effects on energy and the environment. The development of a magnetically levitated transportation system could increase considerably the demand for the metals that might be used in the necessary superconducting or magnetic alloys. Widespread use of nuclear power could alter sharply the consumption patterns of fossil fuels and the related pressures on transportation systems.

The materials cycle can be perturbed in addition by external factors such as legislation. The Clean Air Act of 1970, for example, created a strong new demand for platinum for use in automotive exhaust-cleanup catalysts. The demand may be temporary, since catalysis has been questioned as the best long-term solution to the problem, but whatever platinum is required will have to be imported, in large measure, in the face of a serious trade deficit. Environmental legislation also will require extensive recovery of sulfur from fuels and from smelter and stack gases; by the end of the century, the tonnage recovered annually could be twice the domestic demand. Such repercussions leave little doubt of the need to approach the materials cycle systematically and with caution.

The Materials Revolution

Man historically has employed materials more or less readily available from nature. For centuries he has converted many of them, first by accident and then empirically, to papyrus, glasses, alloys, and other functional states. But in the few decades since about 1900, he has learned increasingly to create radically new materials. Progress in organic polymers for plastics and rubbers, in semiconductors for electronics devices, in strong, light-weight alloys for structural use has bred entire industries and accelerated the growth of others. Engineers and designers have grown steadily more confident that new materials somehow can be developed, or old ones modified, to meet unusual requirements. Such expectations in the main have been justified, but there are important exceptions. It is by no means certain, for example, that materials can be devised to withstand the intense heat and radiation that would be involved in a power plant based on thermonuclear fusion, although the fusion reaction itself is not primarily a materials problem.

This expanding ability to create radically new materials stems largely from the explosive growth that has occurred during this century

in our scientific understanding of matter. Advances in knowledge also
have contributed much to the unifying ideas of materials science and
engineering -- wave mechanics, phase transitions, structure/property
relationships, dislocation theory, and other concepts that apply to many
classes of traditionally "different" materials. Certain semiconductor
materials are perhaps the archetypal example of the conversion of
fundamental knowledge to materials that meet exacting specifications.
Our basic understanding of most materials, however, falls short of the
level required to design for new uses and environments without considerable experimental effort. Hence, it is important to keep adding to the
store of fundamental knowledge through research, although much empirical
optimization will probably always be needed to deal with the complex
substances of commerce.

The Systems Concept

Thorough systems analysis has been used to a moderate extent in
materials science and engineering, but it must become basic to the field
in view of the complexity of modern materials problems and of the fact
that the materials cycle itself is a vast system. The need for the
systems approach is apparent in the ramifications of replacing copper
wire with aluminum in many communications uses in which the substitution
would not have worked well until a few years ago. The move was triggered
by changing relative prices and supply conditions of the metals. A
research and development program produced aluminum alloys with the
optimum combination of mechanical and electrical properties. The
aluminum wire still had to be somewhat larger in diameter than copper
wire, however. Thus wire-drawing machines had to be redesigned, in part

to avoid residual strain in the aluminum wire. Thicker wire, in addition, requires larger conduits, which take more space. And new joining techniques were necessary to avoid corrosion mechanisms peculiar to the aluminum wire.

Products like nuclear reactors, jet engines, and integrated circuits (Figures 1, 2, 3) are systems of highly interdependent materials, each carefully adapted to its role in the total structure. The reaction of such a system to a breakdown at one point is evident in the intended use of a promising graphite-epoxy composite for the compressor blades of a British engine for an American jet airliner. The material was not developed on schedule, to the required degree of service reliability. The repercussions reached well beyond the resulting redesign of the engine. The respective governments were compelled to extricate both companies involved from financial crises, in an atmosphere of sharp debate over domestic and foreign policy.

Materials in a Changing Context

Materials and the associated science and engineering exist in a social and economic context that has changed markedly in the past five years. A pertinent indicator is the National Colloquy on Materials Science and Engineering, held in April 1969: the proceedings[*] took virtually no notice of the field's close ties to the environment, an omission that could hardly occur today. Materials are involved also in

---

[*] Materials Science and Engineering in the United States, Rustum Roy, Ed., Pennsylvania State University Press, University Park, Pa., 1970.

BOILING WATER REACTOR

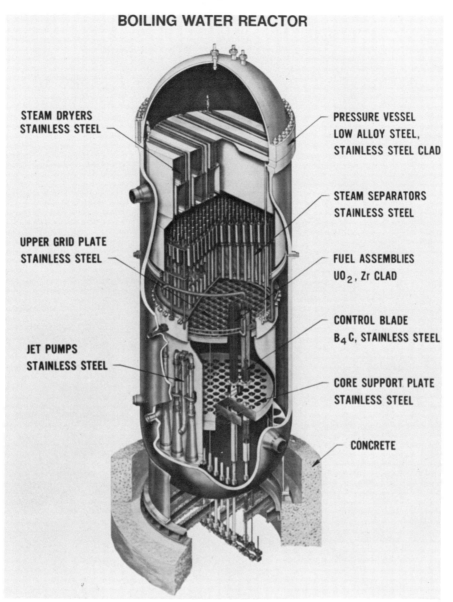

STEAM DRYERS
STAINLESS STEEL

UPPER GRID PLATE
STAINLESS STEEL

JET PUMPS
STAINLESS STEEL

PRESSURE VESSEL
LOW ALLOY STEEL,
STAINLESS STEEL CLAD

STEAM SEPARATORS
STAINLESS STEEL

FUEL ASSEMBLIES
UO_2, Zr CLAD

CONTROL BLADE
B_4C, STAINLESS STEEL

CORE SUPPORT PLATE
STAINLESS STEEL

CONCRETE

FIGURE 1. Materials in a Boiling-Water Nuclear Reactor
Materials shown here in a conventional boiling-water reactor for producing
electric power evolved over years of development. A problem for the future
is the perfection of nuclear-fuel assemblies for a commercial breeder reactor.
The uranium dioxide fuel pellets used in the boiling-water reactor will probably
be replaced by uranium-plutonium dioxide pellets. Working out the relevant
characteristics of the new fuel will occupy many scientists and engineers
for several years. (Illustration courtesy of General Electric Company)

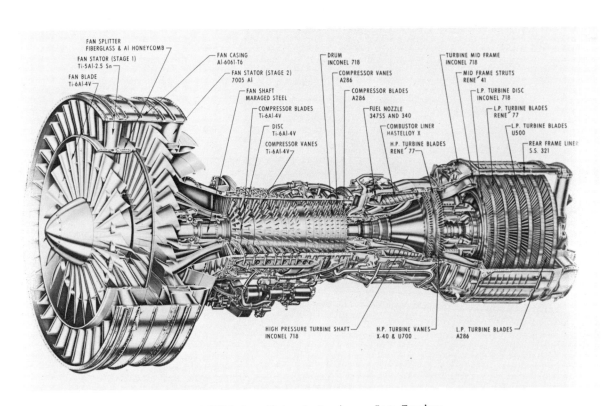

FIGURE 2. Materials in a Jet Engine

Materials complexity is evident in the jet engine, where an overriding goal
is to improve the ratio of thrust to weight. Materials with good potential
for the purpose are composites. Carbon- or boron-reinforced polymers, for
example, might replace the titanium-aluminum-vanadium alloy used in the low-
temperature fan blades (top left). (Illustration courtesy of General Electric
Company)

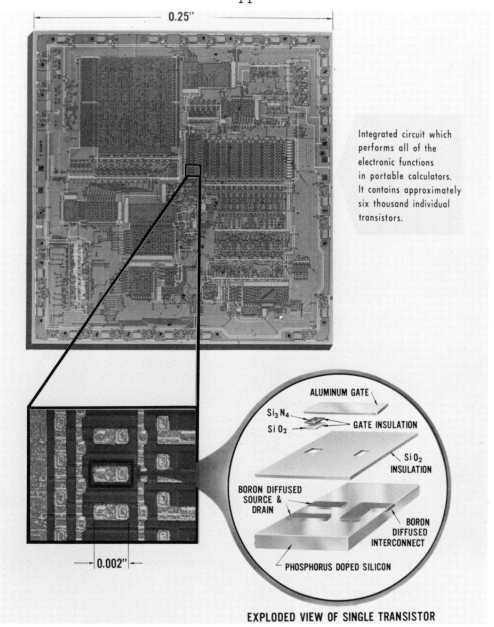

Integrated circuit which performs all of the electronic functions in portable calculators. It contains approximately six thousand individual transistors.

FIGURE 3. Materials in an Integrated Circuit

Among the unusual requirements for semiconducting materials in integrated circuitry and other solid-state electronic devices is the precise processing control of composition and structure in large-scale commercial production of assemblies measured in hundredths and even thousandths of an inch. (Illustration courtesy of Texas Instruments, Inc.)

other kinds of change: the nation's problems with the balance of trade,
federal efforts to stimulate and to assess technology; changing patterns
in spending on basic and applied research and between civilian-oriented
and defense- or space-oriented research and development; and the growing
federal awareness of the importance of materials.

Two fundamental parameters in these matters are population growth
and higher incomes. Between 1900 and 1970, the population of the United
States rose 270 percent, to just under 205 million. For the year 2000
the Bureau of the Census projects a minimum population of 251 million
and a maximum of 300 million. Per-capita gross national product in constant 1958 dollars, meanwhile, has risen steadily, from $1,351 in 1909
to $3,572 in 1971. Both population and per-capita gross national pro-
duct are expected to continue to grow, making ever more urgent the
solution of materials-related problems.

Energy, Environment

The emergence of energy as a national problem of the first rank
was reflected in mid-1973 in the President's establishment of a White
House Energy Policy Office and his call for drastically increased federal spending on energy research and development. At the same time
the President asked Congress to authorize a Cabinet-level Department of
Energy and Natural Resources and the splitting of the Atomic Energy
Commission into an Energy Research and Development Administration and
a Nuclear Energy Commission. The energy problem also is reflected in
the formation of the Electric Power Research Institute by public and
private utilities that account for about 80 percent of the nation's
generating capacity. The Institute will supervise research and

development for the electric utility industry and plans to spend some $100 million in 1974, its first year of full operation. The Institute will be funded by self-assessment of member companies and also will seek to work with the federal government and equipment manufacturers.

National concern for the environment has been recognized in the past few years by extensive federal legislation and by the creation of the Environmental Protection Agency and the Council on Environmental Quality. Environmental matters achieved international status with the Stockholm Conference on the Human Environment, held in mid-1972 under the aegis of the United Nations General Assembly. The status was formalized in December 1972 when the General Assembly established a new unit, the U.N. Environmental Programme.

The U. S. Trade Balance

Materials are important factors in this country's balance of trade. The National Commission on Materials Policy has reported that, in 1972, the United States imported $14 billion worth of minerals (including petroleum) and exported $8 billion worth, for a net deficit of $6 billion. If the trends of the past 20 years persist, the Commission said, the deficit could top $100 billion annually by the year 2000. In 1970, the United States imported all its primary supplies of chromite, columbium, mica, rutile, tantalum, and tin; more than 90 percent of its aluminum, antimony, cobalt, manganese, and platinum; more than half of its asbestos, beryl, cadmium, fluorspar, nickel, and zinc; and more than a third of its iron ore, lead, and mercury. Certain science-intensive materials, on the other hand, including organic chemicals and plastics and resins, have produced, consistently, a positive balance of trade (Table 1).

The country's balance of trade has suffered from growing imports of manufactured products (Table 1). This has happened particularly with low-technology (experience-intensive) goods and, to a lesser extent, with high-technology (science-intensive) goods (even allowing for a degree of controversy over which is which). It appears, in fact, that the United States lost its technological leadership in some product areas, although cause-and-effect relationships among research and development budgets, technological initiative, and foreign trade are difficult to establish clearly (and lie, in any case, beyond the competence of COSMAT).

The federal government has initiated modest efforts to stimulate civilian research, development, and innovation, so as to help recover technological initiative (which may have been lost, for example, in steel and titanium). The goal is to make U. S. products more competitive at home and abroad, and much of the emphasis will be on manufacturing technology, including materials shaping, forming, assembly, and finishing. The federal efforts include the Experimental Technology Incentives Program of the National Bureau of Standards in the Department of Commerce and the Experimental R&D Incentives Program of the National Science Foundation.

Technology Assessment

Technology assessment has long been practiced, in varying degree, in both industry and government, but a formal federal apparatus was established only recently by the Technology Assessment Act of 1972.

TABLE 1

U.S. Trade Balance in Illustrative Product Categories

(millions)

	1960	1965	1970
Aircraft and Parts	$1,187	$1,226	$2,771
Electronic Computers and Parts	44	219	1,044
Organic Chemicals	228	509	715
Plastic Materials and Resins	304	384	530
Scientific Instruments and Parts	109	245	407
Air Conditioning and Refrigeration Equipment	135	207	374
Medical and Pharmaceutical Products	191	198	333
Rubber Manufacture	108	119	-28
Textile Machinery	104	54	-37
Copper Metal	-62	-132	-171
Phonographs and Sound Reproduction	15	-36	-301
Paper and Paper Products	-501	-481	-464
Footwear	-138	-151	-619
TV's and Radios	-66	-163	-717
Iron and Steel	163	-605	-762
Petroleum Products	-120	-464	-852
Textiles and Apparel	-392	-757	-1,542
Automotive Products	642	972	-2,039

Source: U.S. Department of Commerce.

The Office of Technology Assessment and other mechanisms created by the Act are designed to give the Congress a stronger in-house grasp of the relative merits and side effects of alternative technologies. The Act did not establish a formal technology-assessment function in the Executive Branch. The birth of the Office of Technology Assessment appears nevertheless to be stimulating similar efforts in various parts of the Executive Branch.

Trends in Basic and Applied Research

The relative economic austerity of the past few years has been felt in basic and applied research in both industry and government. Nonfederal spending on research and development has virtually leveled off (in constant dollars), while federal spending has been declining (Figure 4). In current dollars, total federal spending on research and development has been rising slowly since 1970, but the emphasis has been shifting away from defense and space toward civilian-oriented areas (Figure 5). Expenditures on space have been falling, while spending on domestic programs has been rising slightly faster than on defense research and development (although starting from a much smaller base). In constant dollars, federal spending on both basic and applied research leveled off in the late 1960's; more recently, spending on basic research has declined slightly, while that on applied research has risen slightly (Figure 6).

The Federal Approach to Materials

The federal government has not yet developed a comprehensive national policy on materials. Materials-related responsibilities are

FIGURE 4

RESEARCH AND DEVELOPMENT SPENDING
IN THE UNITED STATES 1953-72

AVERAGE ANNUAL RATE OF GROWTH %						
	CURRENT $			CONSTANT $		
YEAR	TOTAL	FEDERAL	NON-FEDERAL	TOTAL	FEDERAL	NON-FEDERAL
1953-61	13.7	16.3	10.1	11.3	13.9	7.8
1961-67	8.4	7.7	9.7	6.3	5.6	7.5
1967-72	3.4	1.0	6.8	-1.0	-3.3	2.3

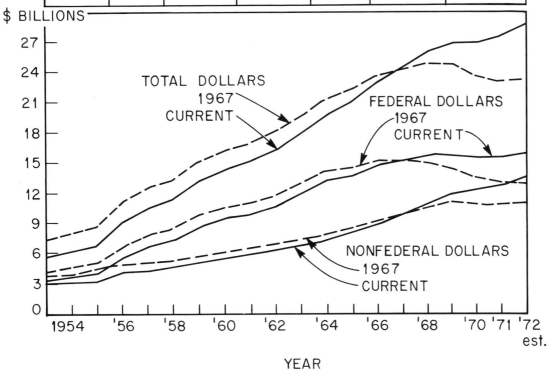

SOURCE: NATIONAL SCIENCE FOUNDATION (1972)

FIGURE 5

CONDUCT OF FEDERAL RESEARCH AND DEVELOPMENT
(OBLIGATIONS CURRENT DOLLARS)

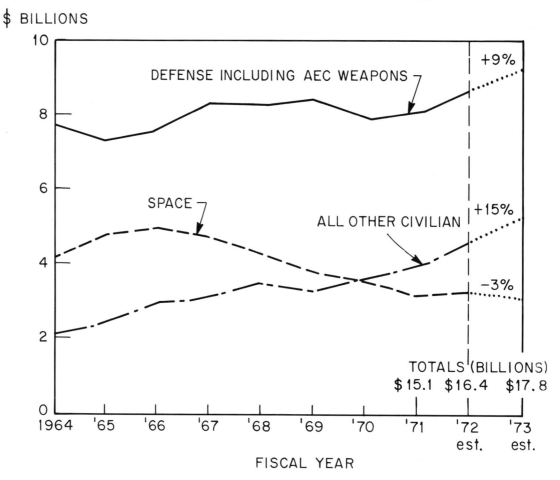

SOURCE: OFFICE OF SCIENCE AND TECHNOLOGY (1972)

FIGURE 6

TRENDS IN FEDERAL BASIC AND APPLIED RESEARCH

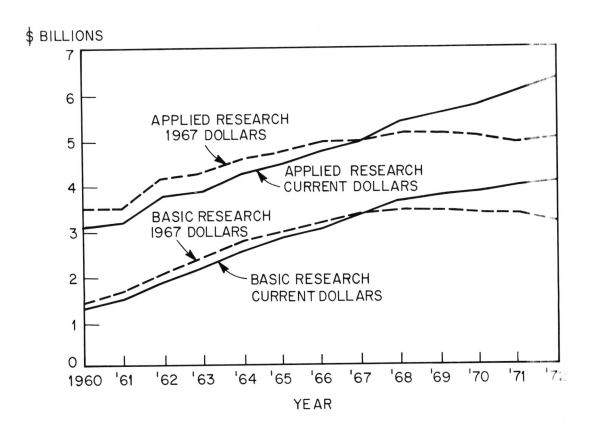

SOURCE: NATIONAL SCIENCE FOUNDATION (1972)

diffused among a variety of formal and ad hoc committees and advisory groups, such as the Interagency Council for Materials. The government is assisted also by groups like the National Materials Advisory Board and the Committee on Solid State Sciences in the National Research Council. The gradual emergence of a more coherent federal approach to materials questions, however, would appear to be implicit in certain developments of the past few years.

The Resource Recovery Act of 1970 created a National Commission on Materials Policy, whose charge was "to enhance environmental quality and conserve materials by developing a national materials policy to utilize present resources and technology more efficiently, to anticipate the future materials requirements of the nation and the world, and to make recommendations on the supply, use, recovery, and disposal of materials." The Commission reported to the President and to the Congress in June 1973.

The Mining and Minerals Policy Act of 1970 requires the Department of the Interior to make annual reports and recommendations for action in relation to a national minerals policy. The Second Annual Report under this Act was published in June 1973.

The creation of a Materials Research Division in the National Science Foundation brought into clearer focus the existence of a multi-disciplinary materials-research community.

Finally, recent years have seen considerable interest in the idea that the earth's finite content of resources for industrial materials (including fuels) restricts severely the industrial growth

that traditionally has been considered the basis of economic and
societal health.[*] This concept of "limits to growth," and the related
idea of a "steady-state society," are not within the scope of this
study. They are, however, further indications of the changing context
in which materials science and engineering exists and in which, we
believe, the field has useful contributions to make.

---

[*] See, for example, H. Brooks, "Materials in a Steady State World,"
Metallurgical Transactions, 3, 759 (1972).

SCIENCE AND ENGINEERING IN MATERIALS ACTIVITIES

Nature and Style of Materials Science and Engineering

In its two decades as a discernibly evolving field, materials science and engineering has reflected growing awareness of the central role of materials in society and has experienced increasingly stringent demands imposed on materials by complex technologies. The significance of the field has been reinforced in the past few years by public concern over the quality of the environment and the availability of natural resources.

Activities in materials science and engineering typically range from basic, curiosity-motivated research to applications-directed development. In its most ambitious reaches, the field relates fundamental understanding of the behavior of electrons, atoms, and molecules to the performance of devices, machines, and structures. It links basic research to the solution of practical problems. At the same time, materials scientists and engineers rely heavily on empirical (experienced-based) knowledge. One instance is the management of the pervasive problem of stress-corrosion cracking, which, for example, can cause catastrophic failure in vessels and piping in the electric-power and other industries. This complex problem is still inadequately understood in fundamental terms. Often it can be avoided, nevertheless, by the use of experience-based knowledge in materials selection and design.

The application of basic and empirical knowledge to function is exemplified by the transistor. The device grew out of a practical need, foreseen in the mid-1930's, to bypass certain intrinsic limitations of vacuum tubes and relays in communication systems. Various approaches to the problem were examined and abandoned. The search narrowed eventually to solid materials, where a few scientists settled on a hunch that the answer lay in semiconductors. To move from the hunch to the transistor, however, required intensive basic research on the behavior of semiconducting materials. It required also the semiempirical development of zone refining and other techniques to make virtually perfect single crystals of silicon of unprecedented purity.

This close linkage of knowledge to function is characteristic of the achievements of materials science and engineering. Selected examples appear in Table 2. A second characteristic of the field is that the initiative for new materials developments, and for the attendant basic research, springs most often from a practical problem, however dimly perceived. It is true that fundamental work on materials has turned up unexpected, momentous discoveries, such as high-field superconductors. But more frequently, the basic studies have been stimulated by a discovery or invention whose exploitation required greatly expanded fundamental research. Thus the tunnel diode and the laser largely preceded and spurred the extensive basic work on the tunnel and laser effects in materials.

Multidisciplinarity and Interdisciplinarity

The bodies of knowledge required for progress in materials, and particularly for solving complex technological problems, often do not coincide with those of the traditional disciplines. Materials science and engineering, as a result, has come to embrace a number of traditional disciplines and segments of disciplines (Figure 7), and

TABLE 2

Selected Achievements in Materials Science and Engineering

Examples of Applications	Material or Process	Basic Research
Transistor, integrated circuits, tunnel diodes, impatt diodes, charge-coupled devices.	Zone refining, float-zone crystal growth, controlled doping in Czochralski growth, epitaxial growth, controlled alloying, diffusion, oxide masking, photo- and electron-beam lithography.	Elemental semiconductors, effects of impurities on conduction properties, impurity chemistry (segregation, alloy systems), crystal-growth studies, dislocations, surface chemistry.
Abrasives.	Synthesis of diamond. Boron nitride.	Phase equilibria studies under extremes of pressure and temperature.
Superconducting solenoids for high magnetic fields. Ultrasensitive electromagnetic signal detectors. Cryogenic logic.	New superconductors: high transition temperature, high critical current (e.g., β-tungstens). Superconducting switches. New effects - Josephson effect - in thin superconducting films.	Superconductivity. Electrical, magnetic, and thermodynamic properties of metals at extremely low temperatures. Many-body theory. Lattice modes.
Cheaper plate glass.	Float-glass process.	
Joining techniques. Scotch tape. Band-aides. Epoxy cements.	Structural adhesvies. Pressure-sensitive adhesives. Anaerobic adhesives.	Rheology. Physical chemistry of surfaces. Synthesis of compounds.
Cheaper steelmaking. Longer-life furnace linings.	Chemistry of steelmaking. Basic oxygen process.	High-temperature phase equilibria.
Ovenware.	Glass-ceramics.	Thermal expansion of ceramics. Nucleation and phase separation.
Aerospace alloys. Aluminum conductor cables. Copper conductors, electrical contacts. High-strength and magnetic alloys.	Dispersion alloys: Internally oxidized particles to strengthen materials; thoria-dispersed nickel; dispersion-hardened aluminum, copper, and silver. Spinodally decomposed alloys.	Thermodynamics of phase diagrams, chemical processes. Particle strengthening.

TABLE 2

Selected Achievements in Materials Science and Engineering (con't.)

Examples of Applications	Material or Process	Basic Research
Aerospace. Turbine blades. Razor blades. Quality cutlery. Magneto-resistance devices. High-strength magnetic alloys. Spring metals. Heat-shrinkable metals.	Directional solidification, continuous casting, amorphous metals, precipitation-hardened alloys, and sheet rolling.	Solidification studies with transparent analogs. Texture and deformation studies. Heat-treatment effects: precipitation, recrystallization, superplasticity.
Optical communications. Ranging for ordnance and surveying. Machining.	Optically-pumped lasers.	Spectroscopy of impurities in crystalline hosts.
Synthetic textiles.	Spinning of fibers from melts and solutions: rayon, nylon, acrylics, polyesters.	Orientation of macromolecular chains.
Tires.	Vulcanization.	Role of molecular networks in determining the properties of rubber.

In these examples of materials science and engineering at work, note that the flow of events is not necessarily, or even usually, from basic research to application. The transistor grew from a perceived need, followed by basic research. The float-glass process was developed originally without benefit of basic research. Vulcanized tires were in use long before molecular networks became accessible to study, but the research that came later has made vulcanization a much more effective process.

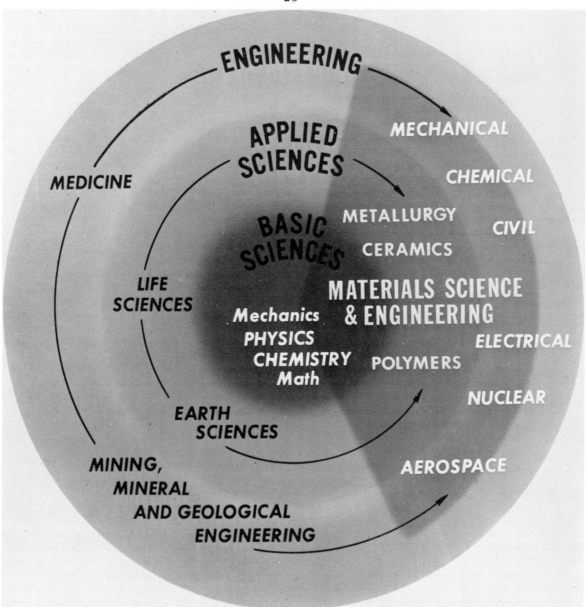

FIGURE 7. Disciplinary Mix in Materials Science and Engineering
Subjects within the shaded sector above are considered to be in the field of
materials science and engineering. Subjects partly or wholly outside the
sector are involved in the field to varying degrees. COSMAT estimates, for
example, that among the 150,000 chemists in the country, there are the equivalent of 50,000 chemists working full time in materials. (Illustration
adapted from Mineral Science and Technology: Non-metallic Materials,
National Academy of Sciences, Washington, D. C. 1969, page 12.)

provides a framework in which the constituent disciplines can assess various opportunities to advance knowledge and help solve societal problems. Together with conventional discipline-oriented activities, the field encompasses both multidisciplinary and interdisciplinary work. In multidisciplinary efforts, scientists and engineers from different disciplines tend to work independently, but readily consult among themselves to benefit from cross-fertilization. In interdisciplinary efforts, two or more individuals from different disciplines collaborate closely on problems or missions that do not fit into single disciplines.

The notion of interdisciplinarity has become increasingly important in the past three decades in materials research and development and in other fields, such as the environmental sciences. Nonetheless, interdisciplinary research remains somewhat contradictory to the scientific tradition that has evolved in the universities. The academic structure -- promotion policies, funding mechanisms, peer-group recognition -- tends to be geared to the individual investigator, working primarily in a single discipline. The fostering of interdisciplinary academic programs in materials, in fact, was one of the goals of the 17 interdisciplinary laboratories established at universities in the early 1960's and funded by the Advanced Research Projects Agency, the Atomic Energy Commission, and the National Aeronautics and Space Administration (see page 37).

The interdisciplinary spirit is more evident in industrial and other mission-oriented organizations. Even there, however, the effectiveness of the mechanism is sensitive to the manner in which the laboratory is organized and managed. It is especially dependent on the steps taken to ease communication among individuals, to create an atmosphere in which people from different disciplines can recognize their need for each other's expertise and the advantages of working together on programs of common purpose.

The transition from research to development also requires care. Early in a project an interdisciplinary research and development group may include mainly basic research people with a few engineers. As the work progresses toward application, more engineers may join the group while some of the basic scientists move on to other programs. This flexible, evolutionary process helps to combat the "not invented here" syndrome that can afflict programs in which research and development are done in sequential steps by different groups.

The establishment of an institutional focus or mission appears to be especially important to laboratories in materials science and engineering. The mission must be carefully chosen and stated, and it must transcend the aspirations of individuals. Themes like communications, energy, and transportation have proved broad enough in some laboratories to draw on many disciplines and yet are specific enough to give the interacting scientists and engineers a sense of common purpose, even in their long-range research. Shorter-range development and engineering problems, in particular, are seldom solved by individuals working on self-chosen bits and pieces related to the problem but not germane to its solution.

National and Institutional Capability

Materials activities are clearly sizable in this country, where 6 percent of the world's population accounts for somewhere between a quarter and a half of the world's annual consumption of natural resources. The United States is very strong in materials science and engineering, but certain weaknesses, if unattended to, could progressively erode the nation's ability to meet the materials needs of its people. These weaknesses are due in part to the diffusion of responsibility for materials plans and programs at the federal level. To a considerable degree, the same diffusion of responsibility is found in the universities, in both education and research. Contributing also to weaknesses in materials are shortcomings in the generation and application of basic knowledge.

National capability in materials science and engineering relies on the trained manpower and basic knowledge produced by the universities and on the application of basic knowledge by industry and other mission-oriented institutions. An organization is better able to assess and exploit new knowledge generated elsewhere when it is able itself to generate new knowledge. Thus knowledge moves more readily from the universities to industry when companies do an appropriate amount of well-chosen basic research. It moves more efficiently also when universities conduct an appropriate amount of applied research. Current difficulties on both scores are pointed out under Universities (page 37) and Industry (page 41).

The importance of materials suggests that materials science and engineering should be a prolific producer of knowledge. That this is so is indicated by the literature as abstracted in Chemical Abstracts. In 1970 Chemical Abstracts abstracted 276,674 papers and patents, of which 45 percent were in materials science and engineering. Over the past two decades, the world-wide literature in materials science and engineering has maintained an annual growth rate of 9 percent whereas the annual growth rate for <u>Chemical Abstracts</u> as a whole has dropped from 8.8 percent in 1950-60 to 6.7 percent in 1960-70. Materials literature originating in the United States has been growing in recent years at 11 percent annually as compared with 13 percent for the Soviet Union, which overtook the United States in materials publications as far back as 1957. The United States produced about 25 percent of the materials papers in 1970; the Soviet Union 33 percent; and Japan 5.8 percent. In the United States, educational institutions were the chief source (50 percent) of the materials literature, followed by industry (25 percent) and government (15 percent). The United States accounted for 40 percent of the world's patents in 1970, and Japan 12.9 percent.

Manpower

Existing data on scientific and engineering manpower generally are not categorized along the multidisciplinary lines of materials science and engineering. We have used a list of specialties characterizing the field, therefore, to extract manpower data from prime sources. On this basis it appears that materials science and engineering involves some 500,000 of the 1.8 million scientists and engineers in the United States. We estimate (Table 3) that there is a full-time equivalent of 315,000 scientists and engineers in the field, including about 115,000 full-time practitioners. Within the latter group are approximately 50,000 professionals holding materials-designated degrees. Engineers, even without counting the materials-designated professionals, constitute the largest manpower group in materials science and engineering; they number 400,000 individuals, and constitute a full-time equivalent of

TABLE 3

Estimates of Manpower in Principal Disciplinary Sectors
of Materials Science and Engineering

Discipline	Total Manpower	Full-Time Equivalent MSE Manpower	
		Total	Doctorates
Chemists	150,000	50,000 (16%)	19,000 (51%)
Physicists	45,000	15,000 (5%)	8,000 (22%)
Metallurgists	40,000	40,000 (13%)	5,000 (13%)
Ceramists	10,000	10,000 (3%)	1,000 (3%)
Other Engineers	1,200,000	200,000[b] (63%)	4,000 (11%)
	1,445,000[a]	315,000 (100%)	37,000 (100%)

---

[a] The total number of scientists and engineers in the United States is about
1.8 million.

[b] Approximately 400,000 engineers are involved significantly in materials
science and engineering. We estimate, conservatively, that they divide
their efforts equally between materials and other engineering activities
and thus are equivalent to 200,000 engineers working full time in
materials.

200,000. The situation with respect to women and minority groups in the materials field appears to be no different from that in science and engineering generally.

The current state of manpower data for materials science and engineering, and our knowledge of the relevant patterns of manpower flow, do not permit reasonable comparisons of the field with the traditional disciplines. However, as the role of materials science and engineering in meeting societal needs becomes more widely understood, it is quite possible that there will be an increasing demand for scientists and engineers in the materials field.

It should be emphasized that the boundaries of materials science and engineering are blurred and continually evolving. The central disciplines and subdisciplines include solid-state physics and chemistry, polymer physics and chemistry, metallurgy, ceramics, and portions of many engineering disciplines. In a broad sense the field also includes segments of mechanics; of organic, physical, analytical, and inorganic chemistry; and of chemical, mechanical, electrical, electronic, civil, environmental, aeronautical, nuclear, and industrial engineering (Table 4).

Government

Materials science and engineering has been shaped in a major way in the past two decades by federal research and development programs that evolved in response to national needs and goals. Direct federal funding of materials research and development totaled some $260 million[*] in

---

[*] Other data suggest that the figure may be as high as $300 million, depending on the definition of terms. Some agencies, and COSMAT, consider research in solid-state physics, for example, to be materials research, while others do not.

TABLE 4

Distribution of Materials Scientists and Engineers
by Category of Activity

Category	% of Professionals in Category Who Are in MSE	% of Total MSE[a]
FROM THE SCIENCE REGISTER		
Polymer and Organic Chemistry	51	6.7
Physical Chemistry	76	3.4
Analytical Chemistry	60	2.6
Solid-State Physics	93	2.0
Inorganic Chemistry	85	1.8
Other Physics	17	1.6
Other Chemistry	15	1.5
Atomic and Molecular Physics	96	0.7
Optics	38	0.5
Earth Sciences	2	0.2
		21%
FROM THE ENGINEERS REGISTER		
Structural Engineering	42	12.6
Metallurgical Engineering	100	11.0
Electromagnetic Engineering	42	10.2
Chemical Engineering	92	9.5
Work Management and Evaluation	18	8.8
Dynamics and Mechanics	40	7.7
Engineering Processes	60	5.4
Heat, Light, and Applied Physics	75	5.4
Automation & Control Instrumentation	45	4.7
Ceramic Engineering	100	1.9
Information and Mathematics	20	1.8
Other Engineering	30	0.1
		79%
		100%

[a] The distributions between the science and engineering portions of this
listing have been adjusted to 21 percent and 79 percent, respectively, in
accordance with the physics plus chemistry percentages shown in Table 3.

Source: 1968 National Register of Scientific and Technical Personnel
(National Science Foundation) and 1969 National Engineers Register
(Engineering Manpower Commission).

fiscal 1971, according to the Interagency Council for Materials; in constant dollars this figure is about equivalent to $185 million spent in fiscal 1962. (Indirect federal funding of materials research development through hardware contracts is estimated at least to equal direct funding, giving a total of some $0.5 billion in federal materials research and development in 1971). The funding is widely scattered (Tables 5, 6). Only one agency, the National Science Foundation, has an identified mandate to support science and technology generally. All the others, commanding 90 percent of the budget, support the science and technology related to their missions.

Governmental materials laboratories, which received about a third of the federal research and development funds in materials in 1971, concentrate primarily on identified, mission-oriented problems. To support this work they do exploratory research at a level that appears to vary from laboratory to laboratory in the range of 5 to 15 percent of their total funding. Some federal laboratories have become centers of excellence in specific areas. These include the Air Force Materials Laboratory, Wright-Patterson Air Force Base, Dayton, Ohio, in composite materials; the Atomic Energy Commission's Oak Ridge National Laboratory in radiation damage and neutron diffraction; and the National Bureau of Standards in polymeric materials.

The federal government also operates the National Standard Reference Data System, which is administered and coordinated by the National Bureau of Standards. This program provides critically evaluated numerical data on the physical and chemical properties of well-characterized substances and systems. The Bureau, in addition, operates the Standard Reference Materials program, which now can provide standard samples of more than 800 materials.

TABLE 5

Direct Federal Funding of Materials Research and Development by
Agency, Type of Research, and Performer
(Fiscal year 1971; millions)

Agency	Type			Performer					Total
	Basic Research	Applied Research	Experimental Development	University	Federal Contract Research Centers	Other Non-Profit	Industry	Government In-House	
Agriculture	$ 9.6	$ 11.8	$ 1.3	$ 0	$ 0	$ 0	$ 0	$ 22.6	$ 22.6
AEC	39.3	42.9	0	18.5	57.7	2.5	2.6	0.8	82.2
NBS	2.1	8.4	0	10.7	0	0	0	10.5	10.5
ARPA	14.0	6.0	0	10.7	0	0.7	6.6	1.9	20.0
Army	3.3	13.9	0	1.7	0	0.3	2.3	12.9	17.2
Navy	10.3	9.7	4.4	2.6	0	1.2	5.1	15.4	24.4
Air Force	3.9	23.4	11.5	2.5	0	0.3	26.6	9.4	38.8
HEW	1.4	2.2	0	1.4	0	1.2	0.9	0	3.6
HUD	0	0	0	0	0	0	0	0	0
Interior	0.1	3.4	0	0.3	0	0	1.8	1.3	3.5
NASA	4.7	14.7	3.2	0.8	0	0.8	3.9	17.0	22.6
NSF	8.9	1.7	0	10.7	0	0	N.A.	0	10.6
DOT	0	2.0	2.2	3.0	1.0	0	0	0.5	4.2
Totals[a]	$ 97.6	$ 140.0	$ 22.6	$ 52.2	$ 58.7	$ 7.1	$ 49.9	$ 92.4	$260.0

——— Total: $260 million ———

——— Total: $260 million ———

Total: $260 million

a   Totals may not add exactly because of rounding in the several compilations used for these figures.

Source: Interagency Council for Materials.

Note: Other data suggest that the $260 million total shown above may be as high as $300 million and the University total as high as $75 million, depending on definition of terms. Some agencies, and COSMAT, consider research in solid-state physics to be materials research, for example, while others do not.

TABLE 6

Direct Federal Funding of Materials Research and Development
by Agency and Field of Materials
(Fiscal year 1971; millions)

Agency	Metallic Materials	Organic Materials	Inorganic Nonmetallic Materials	Composite Materials	Fuels, Lubes, Fluids	Other Materials
Agriculture	$ 0	$ 22.8	$ 0	$ 0	$ 0	$ 0
AEC	38.4	4.7	34.1	3.4	0	1.6
NBS	1.8	1.9	2.5	0	0	4.3
ARPA	8.1	2.2	9.3	0.5	0	0
Army	7.3	3.9	2.0	1.5	2.4	0
Navy	12.5	3.9	4.6	2.2	0.9	0.3
Air Force	12.0	7.3	3.2	14.2	0.5	1.5[b]
HEW	0.6	1.6	0.1	1.5	0	0
HUD	0	0	0	0	0	0
Interior	2.2	1.0	0.3	0	0	0
NASA	12.6	4.2	1.7	2.6	1.5	0
NSF	2.1	0.6	a	a	0	8.1[c]
DOT	0.2	0.2	4.2	0	0	0
Totals[d]	$ 97.6	$ 53.6	$ 61.9	$ 25.9	$ 5.3	$ 15.9

Total: $260 million

a   Not reported under materials research and development

b   Materials physics, solid state

c $7.6 solid state, 0.5 biomaterials

d   Totals may not add exactly because of rounding in the several compilations used for these figures.

Source: Interagency Council for Materials.

Note: Other data suggest that the $260 million total shown above may be as high as $300 million, depending on definitions of terms. Some agencies, and COSMAT, consider research in solid-state physics to be materials research, for example, while others do not.

Universities

Some areas of materials science and engineering, such as metallurgy and ceramics, are full-fledged academic disciplines. Areas represented as formal degree programs include materials science, polymer science, solid-state science, and materials engineering. Despite the increasing definitiveness of such programs, the nature and uses of materials are so broad and pervasive as to continue to require close interaction among many disciplines.

At least half of the identifiable research on materials in universities is done on the 28 campuses where materials research centers have been established in the past decade. Twelve of these were selected in the early 1960's as the sites of interdisciplinary laboratories (IDL's). These laboratories were sponsored by the Advanced Research Projects Agency in the Department of Defense (ARPA) as an experiment in improving the sophistication of materials research and increasing the number of materials specialists. A notable feature of these laboratories has been the availability of block funding for locally selected research programs and central facilities. Responsibility for the interdisciplinary laboratories program was assumed on July 1, 1972, by the Materials Research Division of the National Science Foundation; the laboratories were then renamed Materials Research Laboratories. In the spring of 1973 the Foundation announced plans for two new Materials Research Laboratories. One will focus on the technology of joining, the other on polymers.

The National Aeronautics and Space Administration (NASA) set up three block-funded programs at universities in the 1960's, but at lower levels of support than for the interdisciplinary laboratories. In the same period the Atomic Energy Commission (AEC) established block-funded materials research centers at two universities, one of them already housing an interdisciplinary laboratory. Since the start of these 17 programs set up by ARPA, NASA, and AEC, 11 additional universities have formed analogous materials research centers, mainly on their own initiative. They use the concept of central facilities, but are mostly without block funding.

The COSMAT study found that, typically, these 28 centers ranked high in education and individual basic research, the traditional functions of the university. Most of the centers ranked low in interaction with industry and in innovative methods of operation. Some centers are doing a relatively significant amount of interdisciplinary work, one measure of which is the authorship of the resulting scientific papers. In three quarters of the centers, 10 to 15 percent of the papers covered by our evaluation were written jointly by faculty from two or more departments. This contrasts with an average of 2 percent for papers from materials-designated departments.

Because of the difficulty of obtaining comparable data from different schools, this evaluation cannot be fully accurate and complete. However that may be, universities have produced relatively little so far in the way of new materials per se. An important reason, we believe, is that the academic community traditionally has resisted interdisciplinary and applied research. We have noted already that the reward structure within the university is tilted strongly toward the disciplines; moreover, no funding agencies have clearly rewarded excellence in interdisciplinary activities at universities. All the materials research centers indicated that they plan to shift their emphasis somewhat toward applied research. The area mentioned most often was biomaterials. This field is highly interdisciplinary and intellectually stimulating, but the corresponding body of technology is much smaller than in other areas, such as ceramics, polymers, and electronic materials.

Materials Degrees. Formal undergraduate curricula in materials appear to be confined to materials-designated degree programs, which are located almost entirely in engineering schools. Some 60 programs of this kind are accredited in the country's 250 engineering schools. These and 30 unaccredited programs award annually somewhat more than 900 materials-designated baccalaureate degrees, or about 2 percent of the total engineering baccalaureates conferred annually. Currently, more than half of the materials-designated departments average fewer than 10 baccalaureates per year, a situation which will become increasingly difficult to justify.

About 50 institutions in the United States offer graduate degrees in materials. The 270 materials-designated doctorates awarded in 1971-72 amounted to about 7% of the total engineering doctorates conferred. In solid-state physics, which provides a major component of the professional manpower in materials science and engineering, doctorates awarded annually appear to number about 370. Annual output of materials-research doctorates awarded in chemistry and in nonmaterials-designated engineering programs is at least double the number in solid-state physics. Thus we estimate that the annual output of doctorates in the field of materials is about 1400, but there is much uncertainty

in this figure because of difficulties in the "materials identification."

Research Funding. University research in materials is supported almost

entirely by federal funds at an annual level (in 1971) of about $52.2

million* (Table 5). This was 20 percent of the $260 million** in total,

direct federal support for materials research and development in 1971

and about 3 percent of total federal support for research and develop-

ment in the universities. Some 35 percent of the support for university

research in materials was provided by the Atomic Energy Commission

and about 20 percent each by the National Science Foundation and the

Advanced Research Projects Agency. (In fiscal 1973 the Foundation

estimates its materials research support at $35 million, more than

triple the $10.7 million of 1971, but the major part of the increase

arises from internal regrouping into the materials research category.)

Relatively little identified support is available as yet from agencies

like the Departments of Health, Education, and Welfare; Housing and

Urban Development; and Transportation.

Almost 60 percent of the federal support for university research

in materials goes to the 12 universities where the National Science

Foundation Materials Research Laboratories are located. About 15

percent of the support goes to the 16 additional schools that have

---

* The figure could be as high as $75 million, depending on definitions of
  terms. COSMAT and some agencies, for example, consider research in
  solid-state physics to be materials research, while others do not.

** The figure could be as high as $300 million, for the reason given in
   the preceding footnote.

materials research centers of other kinds. On the 12 campuses where the NSF Materials Research Laboratories are located, slightly more than 25% of materials-research funds are received by materials-designated departments, slightly less than 15% by other engineering departments, and about 60% by physics and chemistry departments.

Industry

Materials are vital to any nation's economy. The consumption of basic materials in the United States has been rising steadily (Table 7), along with the population and standard of living. The resulting drain on resources has drawn sharp attention at various times, but especially in the past few years, as has the impact of waste or residual materials on the environment. From these concerns has emerged a need to think in terms of specific materials flows. One such flow diagram (Figure 8) has been developed to show the interplay of residuals and recycling in relation to environmental quality. Though qualitative, it portrays the materials streams that might be involved in a quantitative model.

We mentioned at the outset that without materials there would be no gross national product. By the same token, analysis of the gross national product is useful in putting materials into perspective. It shows that, broadly considered, about a fifth of this country's gross national product may be said to originate in the extracting, refining, processing, and forming of materials into finished goods other than food and fuel (Table 8). These operations on materials account for perhaps one tenth of the nation's consumption of fuels. A third

TABLE 7

Consumption of Selected[a] Basic Materials in the United States
(Millions of Tons)

	1950[b]	1971[b]		1950[b]	1971[b]
Aluminum	1.3	5.5	Clays	39.5	55.1
Calcium	N.A.	90.3	Gypsum	11.4	15.7
Copper	2.0	2.4	Pumice	0.7	3.5
Iron	94.5	122.2	Sand and gravel	370.9	987.7
Lead	1.4	1.3	Stone, crushed	N.A.	823.0
Magnesium	N.A.	1.1	Stone, dimension	N.A.	1.8
Manganese	1.1	1.2	Talc	0.6	1.1
Phosphorus	1.7	5.1	. . .		
Potassium	1.2	4.5	Agricultural fibers	N.A.	2.1
Sodium	N.A.	19.0	Forest products	N.A.	237.0
Sulfur	6.8	12.4	Plastics	1.0	10.0
Zinc	1.1	1.2			

a   Commodities used in excess of 1 million tons in 1971. Totals include government stockpiling, industry stocks, and exports. Foods and fuels are not included.

b   1950 actual; 1971 estimated.

Source: First annual Report of the Secretary of the Interior under the Mining and Minerals Policy Act of 1970, March 1972. Figures for agricultural fibers, forest products, and plastics compiled by COSMAT from various sources.

43

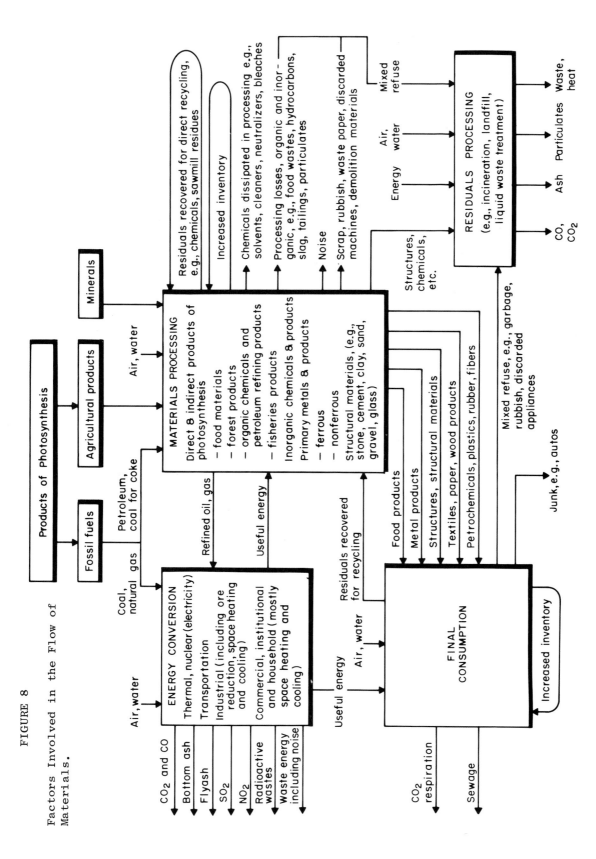

FIGURE 8

Factors Involved in the Flow of Materials.

Source: Robert U. Ayres and Allen V. Kneese, "Pollution and Environmental Quality," Quality of the Urban Environment, Harvey S. Perloff, Ed., Resources for the Future, Inc., Washington, D.C., p. 37.

TABLE 8

Selected Industry Components of the Gross National Product (1971)

(1971 GNP = $1,050,356 million)

	Millions	% of GNP
Metal Mining	$ 1,290	0.12
Mining and Quarrying of Nonmetallic Metals	1,654	0.16
Stone, Clay and Glass Products	8,710	0.83
Primary Metal Industries	18,923	1.80
Fabricated Metal Products	16,427	1.56
Machinery, except Electrical	26,066	2.48
Electrical Machinery	22,388	2.13
Transportation Equipment, except Motor Vehicles	14,582	1.39
Motor Vehicles and Motor Vehicle Equipment	22,824	2.17
Instruments	6,456	0.61
Miscellaneous Manufacturing Industries	4,144	0.39
Chemicals and Allied Products	20,387	1.94
Rubber and Miscellaneous Plastic Products	7,371	0.70
Lumber and Wood Products, except Furniture	6,395	0.61
Furniture and Fixtures	3,984	0.38
Paper and Allied Products	9,357	0.89
Textile Mill Products	8,234	0.78
Apparel and Other Fabricated Textile Products	9,293	0.88
Leather and Leather Products	2,219	0.21
	$210,704	20.03

Source: U.S. Department of Commerce.

measure is manufacturing employment related to materials, which was
just over 16 million in 1970 or about 21 percent of total employment.

Still another view of the economic significance of materials
is provided by an analysis of structural changes in the economy
arising from changes in technology.[*] The analysis was based on input-
output tables for 1947 and 1958, a period in which tonnages and dollar
values of materials were rising steadily. The results show strikingly
the trend toward a service economy during this period, as indicated
by the relative increases in "nonmaterial" or "general" inputs, which
were largely balanced by relative decreases in the inputs of materials
and semifinished goods. The iron and steel sector, for example,
declined relatively some 27 percent although it grew about one third
in absolute terms.[**] Besides the trend toward a service economy,
the decline reflected substitution of aluminum and plastics for steel
as well as weight-reducing design changes based on improvements in the
properties of steel. Nonferrous metals decreased 23 percent, relatively,
as the greater use of aluminum was more than offset by declines in
other nonferrous metals. The analysis showed also that:

> "The classical dominance of single kinds of
> material -- metals, stone, clay and glass, wood,
> natural fibers, rubber, leather, plastics, and
> so on -- in each kind of production has given
> way by 1958 to increasing diversification of

---

[*] Carter, A. P., "The Economics of Technological Change, "Scientific
American, 214, 4, 25 (1966).

[**] In fact, iron and steel output plunged sharply in 1958 in the course
of a brief recession. It was still slightly higher than in 1947,
however, and was growing again by 1959.

the bill of materials consumed by each
industry. This development comes from
interplay between keenly competitive
refinement in the qualities of materials
and design backward from end-use specifications."

Industrial Research and Development. Accurate figures are not available for materials research and development in industry. Data for
industrial research and development in general (Table 9) indicate that
"All Industries" planned a 4 percent increase in spending in 1972
including federally-funded industrial research and development. The
metals-producing industries -- steel, nonferrous metals, fabricated
metals -- were expected to remain essentially level in 1971-72 in
constant dollars. This would have meant an 8 to 10 percent decrease
in research and development actually performed because of rising costs.
Decreases in work performed were also indicated in paper and in stone,
clay, and glass. "All Manufacturing" showed an estimated increase of
only 2 percent in 1972, again amounting to a decrease in research and
development actually performed. Even in high technologies like
aerospace (no change in 1972) and electrical machinery and communications (up to 2 percent), research and development has not kept up with
rising costs. Industrial research and development as a percentage of
sales (Table 10) held level or declined in 1972 in all areas except
aerospace. Federally funded research and development in All Manufacturing (Table 11) is declining, both in dollars and as a percentage
of total industrial research and development.

More recent figures (Table 12) show a brightening picture for
company-funded research and development, although substantial differences exist among individual industries. Spending on basic research

TABLE 9

Industrial Research and Development
(Includes federally funded industrial R&D)

	1970 Actual	Est. 1971	Planned 1972	1975	1971-72	1972-75
			(Millions)			(Percent)
Steel	$ 131	$ 122	$ 132	$ 149	8	13
Nonferrous Metals	134	165	155	234	-6	51
Machinery	1,727	1,831	1,923	2,173	5	13
Electrical Machinery & Communications	4,324	4,410	4,498	5,353	2	19
Aerospace	5,173	4,914	4,914	5,061	0	7
Autos, Trucks & Parts & Other Transportation Equipment	1,475	1,475	1,504	1,609	2	7
Fabricated Metals & Ordnance	183	176	183	210	4	15
Professional & Scientific Instruments	694	756	824	972	9	18
Lumber & Furniture	24	31	36	38	16	6
Chemicals	1,809	1,827	1,882	2,145	3	14
Paper	119	133	133	166	0	25
Rubber Products	238	281	295	336	5	12
Stone, Clay & Glass	188	169	169	198	0	17
Petroleum Products	608	492	522	606	6	16
Food & Beverages	198	208	225	263	8	17
Textile Mill Products & Apparel	64	60	66	81	10	23
Other Manufacturing	98	117	124	161	6	30
ALL MANUFACTURING	17,187	17,167	17,585	19,755	2	12
Nonmanufacturing	669	723	1,063	1,711	47	61
ALL INDUSTRIES	17,856	17,890	18,648	21,466	4	15

Source: National Science Foundation (1972).

TABLE 10

Industrial Research and Development as Percent of Sales[a]

	1970	1971	1972[b]	1975[b]
Steel	.34%	.31%	.28%	.26%
Nonferrous Metals	.76	.90	.79	.93
Electrical Machinery	8.51	8.17	7.72	7.23
Machinery, Other	3.08	3.08	2.96	2.61
Aerospace	19.02	20.05	20.88	17.92
Autos, Trucks & Parts & Other Transportation Equipment	2.73	2.23	2.05	1.71
Stone, Clay & Glass	1.06	.81	.74	.70
Fabricated Metals	.44	.41	.40	.37
Instruments	5.71	6.39	6.27	5.52
Chemicals	3.71	3.54	3.41	3.16
Paper	.74	.51	.47	.45
Rubber	1.36	1.49	1.43	1.34
Petroleum	2.29	1.76	1.73	1.66
Textiles	.29	.26	.26	.24
Food & Beverages	.20	.20	.20	.19
Other Manufacturing	.13	.14	.14	.13
ALL MANUFACTURING	2.63%	2.47%	2.32%	2.08%

[a]   Sales figures are based on company data classified by major product line.

[b]   1972 estimated; 1975 planned.

Source: 1972 McGraw-Hill Survey of Industry Research and Development.

TABLE 11

Federally Financed Industrial Research and Development
(Amounts and percent of total R&D spending by industry)

| INDUSTRY | ------1971------- | | -------1972[b]----- | | ------1975------- | |
	Percent	Million Dollars	Percent	Million Dollars	Percent	Million Dollars
Steel		a		a		a
Nonferrous Metals	5%	$ 8	6%	$ 9	6%	$ 14
Machinery	12	220	10	192	8	174
Electrical Machinery & Communications	50	2,205	48	2,159	42	2,248
Aerospace	80	3,931	76	3,735	72	3,644
Autos, Trucks & Parts & Other Transportation	13	192	12	180	10	161
Fabricated Metals & Ordnance	3	5	3	5	3	6
Professional & Scientific Instruments	25	189	23	190	21	204
Lumber & Furniture		a		a		a
Chemicals	10	183	10	188	11	236
Paper	1	1	1	1	1	2
Rubber Products	15	42	14	41	12	40
Petroleum Products	5	25	5	26	5	30
Food & Beverages	1	2	1	2	1	3
Textile Mill Products & Apparel		a		a		a
Other Manufacturing		a		a		a
ALL MANUFACTURING	41%	$7,008	38%	$6,731	34%	$6,770
Nonmanufacturing	68	492	65	691	60	1,027
ALL INDUSTRIES	42%	$7,500	40%	$7,422	36%	$7,797

[a]   Less than $500,000
[b]   1972 estimated; 1975 planned.
Source: National Science Foundation (1972).

TABLE 12

Company-Funded Industrial Research and Development
(Millions)

	Total R & D			Basic Research		
	1971	1972	1975 (Est.)	1971	1972	1975 (Est.)
All Industries	$10,643	$11,400	$13,950	$ 494	$ 520	$ 650
Drugs & Medicine	505	560	750	95	105	140
Industrial Chemicals	864	890	1,025	100	105	125
Petroleum	488	495	525	22	23	25
Electrical Equipment	2,230	2,400	3,000	109	115	145
Aircraft & Missiles	1,012	975	1,150	34	30	40
All Other	5,544	6,080	7,500	134	142	175

Source: National Science Foundation (1973).

in "All Industries" is projected to rise 25 percent in 1972-75, to
$650 million; spending on research and development overall is expected
to rise 22 percent in the same period, to just under $14 billion. The
source of these figures, the National Science Foundation, notes the
changing nature of industrial basic research. Companies generally
are shifting toward "shorter-term, more relevant, and hence more
economically-justifiable projects."

Need for Research and Development. Industry in this country and abroad
has produced many of the outstanding achievements of materials science
and engineering. They include nylon; the transistor; the high-field
superconductor; the laser; phosphors for television, radar, and
fluorescent lamps; high-strength magnetic alloys; magnetic ferrites;
and polyethylene. These developments occurred in industries that conducted long-range research to expand the basic knowledge on which the
industry ultimately relied. By thus supplementing their experience-
based approach to materials research and development, these industries
established technological leadership for themselves and for their
countries. The resulting cumulative national payoff, though difficult
to measure, is substantial.

Our current shift from aerospace, atomic energy, and defense
toward more civilian-oriented technologies offers industry a wide
variety of fresh technical challenges: in the environment, in energy,
in the quality and safety of consumer goods. Many such challenges will
be met only with the help of sustained basic and applied research. Yet
industry has been cutting back its relatively basic programs in the

past few years. The science-intensive industries have retrenched significantly; the experience-based industries in many cases have virtually eliminated what little basic research they were doing.

Competitive pressures and the cost of research and development are rising steadily. A not-uncommon view is that the penalties of failure in research and development and the liability of high engineering risk have grown too great, while the rewards of success and the achievement of advanced product performance are too easily appropriated by others. Some companies now are reluctant to undertake programs that do not promise to begin to pay for themselves in 5 to 10 years at the most. The payoff period for basic research in materials, in contrast, although it tends to be shorter than in other areas, may sometimes exceed 10 years. A company that is not a technological leader may find that new technology is obtained more sensibly from other companies, by cross-licensing or by royalty agreements. But the company striving to achieve or maintain technical leadership will find a balanced research and development program essential to its success.

More broadly, were basic research in materials science and engineering to be eliminated, the rate of introduction of new technology might not slow noticeably for several years. But then the nation's capability would decline -- precipitously in some high-technology areas. The country could sink to a seriously inferior position internationally in 10 to 20 years. Many industrial managements and, perhaps, the general public are not prepared to wait that long for the fruits of research. But industry should recognize more widely, we believe, that research in materials characteristically has returned good value and that the payoff is more assured than in many other fields. Progress in materials may not depend on public support to the same extent as does progress in astronomy, let us say, but for government, as for industry, materials science and engineering represents a sound investment. As in other fields, the decisions to be made often relate to the appropriate roles of government and private initiative in undertaking research. These can be hard decisions, but they must be made.

NATIONAL CONCERNS AND TECHNICAL CHALLENGES

Today we are faced with growing competition for nonrenewable raw materials and fuels and with low standards of living in much of the world. The latter is an old problem, but it is reemerging in a new setting that features prominently the aspirations of the developing countries, concern for the environment, and the scale of international human activities. These difficulties, in consequence, are attracting more and more attention, both in the U. S. and abroad, shifting to a degree the emphasis on national defense and political prestige toward more civilian-oriented goals and concerns.

Materials science and engineering can help meet the technical challenges of these growing concerns. The relationship between materials research and development and one such concern, health services, is shown in a general way in the partial "relevance tree" of Figure 9. The existing and potential utility of materials research and development in solving a range of other real-life problems will be evident in our discussion of opportunities in materials research (page 97). In addition, we have examined the diverse technical challenges of the country's current concerns from two vantage points: challenges in the materials cycle, and challenges in specific areas of national concern, including a priority analysis based on questionnaire replies.

Challenges in the Materials Cycle

Materials science and engineering, by providing options at the various stages in the total materials cycle, can exert direct, if not

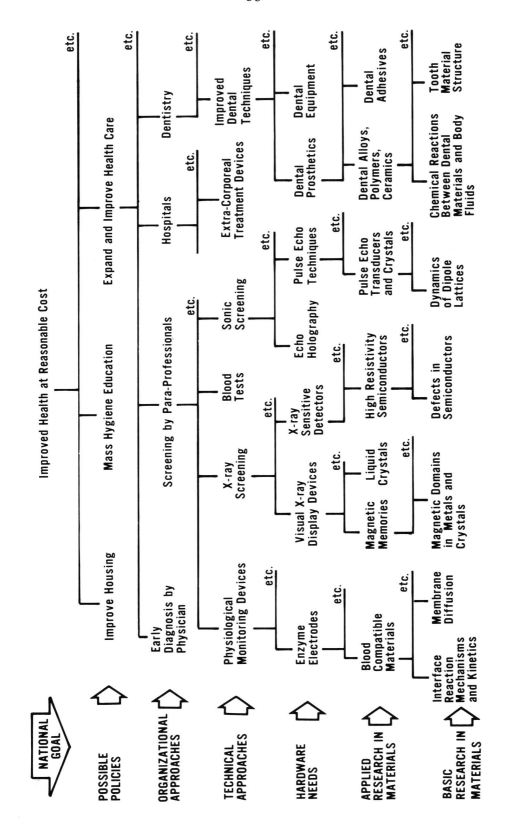

FIGURE 9

PARTIAL RELEVANCE TREE FOR HEALTH SERVICES

The relevance tree shown here for health services and materials is not comprehensive, but illustrates the use of the technique for relating national or other broad goals to pertinent needs in research. Additional pathways between basic research and broad goals have been omitted for the sake of simplicity. Also omitted is the time scale: research on materials, for example, might not begin feeding information upward for 10 years or more.

always immediately visible, effects in the problem areas reflected by national concerns. It can help to slow and sometimes to halt the growth in demand for certain raw materials and fuels. It can help to move hardware technologies in directions that raise living standards at home and abroad. It can help to reduce deleterious effects on the environment to acceptable levels. And it can help to achieve these goals in a manner consistent with a sound U. S. balance of trade.

Exploration. The sensing, information-processing, and transmitting functions of orbiting earth-resources satellites and lunar rovers were made possible by progress in development of electronic and structural materials. Comparable technology could be developed for exploring the ocean floor. For more traditional types of prospecting, instrumental methods should progress rapidly as more is learned of the "signatures" of complex natural materials.

Mining. Ores and minerals, in the future, probably will have to be mined in more hostile environments at less accessible sites. (Manganese and other metals, as well as phosphates, for example, are available on the ocean floor.) Working conditions often may be impossible for human operators. To tap the resources available from ultradeep mines or below the ocean floor will require a new technology, "robotics." In essense, robotics will involve solid-state electronic sensing and information-processing equipment coupled to servomechanical mechanisms that can operate under extreme conditions. The advent of novel equipment of this kind likewise will benefit conventional mining operations. Plasma and rocket-nozzle technology, for instance, has proved useful in drilling the hard, iron-bearing taconite -- which has largely succeeded the heavily depleted high-grade domestic iron ore that was long the mainstay of the nation's steel industry.

Extraction. We need very much to find new means of extracting basic materials from ores of progressively lower grade and from low-grade wastes, processes that are more efficient, that cost less, consume less energy, and cause less pollution. Aluminum already is being extracted from the abundant anorthosite (in the Soviet Union) as opposed to the conventional source, the high-grade but less plentiful bauxite. Under development in the United States are two new aluminum processes: one reduces by about a third the energy required to produce aluminum from alumina by electrolysis; the other produces aluminum in several (nonelectrolytic) steps, starting with various sources of the metal -- not only bauxite, but low-grade alumina-bearing minerals and even clay. The large piles of blast-furnace and open-hearth slag in the Mid-west are potential sources of manganese and phosphate. Longer-range possibilities include simultaneous extraction -- perhaps at very high temperature -- of several materials from "ores" like granite, which contains all the elements necessary to a modern industrial society. For higher-value materials, study seems warranted on electrostatic, electrophoretic, and other novel methods of separation.

Renewable Resources. Considerable scope exists for expanding the range of materials obtained from renewable resources. Wood and vegetable fibers might become important sources of primary organic chemicals, although they are not economically competitive today. Means of "cracking" the lignin molecule, the binding material in trees, could make organic chemicals available from about 25 million tons of lignin disposed of annually in this country in wood wastes with only minor recovery of values. The utility of renewable resources in general might be extended by a variety of methods: better chemical means of recovering basic materials; control of physical properties by chemical or radiation treatment; genetic modification during growth; new ways to make composite materials of natural products; and improved methods of protecting and preserving structural materials made of natural products.

Resource Substitution. The substitution of plentiful for less-plentiful resources is likely to become an especially important task for materials science and engineering in the future. A material may be substituted for another of the same class, as when aluminum replaces copper in electrical conductors, or for one of a different class, as when polyethylene replaces galvanized steel in buckets. We will need substitutes for certain metals that have unique and important properties but threaten to become critically scarce in the not-so-distant future. These include gold, mercury, and palladium. The nation's balance of trade would benefit from substituting manganese for nickel as a stabilizer in stainless steels and substituting domestic ilmenite for imported rutile as a source of titanium.

Even metals and alloys used widely in structural applications may offer broad scope for substitution by other alloys or ceramics based on substances more abundant in nature. The most common substance in the earth's crust is silicon dioxide. It is a basic constituent of glasses, which are remarkably versatile materials used hardly at all in proportion to their potential abundance. The properties of glass include excellent corrosion resistance and very high intrinsic strength. Aluminum and magnesium -- though the energy cost of obtaining them is relatively high -- are plentiful and display useful properties. These include, especially, the high ratios of strength to weight so important in engineering applications.

Processing, Manufacturing. Widespread opportunity exists for new processing and manufacturing techniques that waste less material and use less energy than do current methods. More processes are needed that lead directly from liquids and powders to finished shapes, thereby avoiding, for metals, the ingot and hot-working states. Such processes in general cost less and consume less energy than do the cold-forming and machining required to shape bulk solids. Industry already shapes liquid or powders in many cases: manufacture of float glass, slip casting or compacting of intricate shapes, die casting and plastic molding, and hot forging of sintered metals.

Continuous on-line assembly with minimum human intervention, a continuing objective for production lines, is virtually achieved in the manufacture of integrated circuits, where relatively few of the 200 or more processing steps are controlled actively by operators. The approach should be extended to other areas of processing and manufacturing. Some of the greatest savings in production costs and resources probably will result in the long run from greater use of small, on-line computers and robots. This form of robotics mentioned earlier for mining calls for the imaginative exploitation of a variety of sensing and monitoring devices coupled through minicomputers to control mechanisms.

Environmental Effects. The need to preserve the environment requires continuing development of industrial processes that release fewer harmful effluents or whose effluents can be captured and converted to harmless and preferably useful forms. Some such processes are used widely now. One is the recovery of sulfur from petroleum refinery off-gases. Another is the recycling of the hydrochloric acid that has been displacing the nonrecyclable sulfuric acid in the pickling of steel for cold-forming. The heavy, hard-rubber cases on automobile storage batteries are not reused and often are disposed of by burning; a lighter-weight, reusable plastic case would seem feasible. The metallic salts in polyvinyl chloride film may become an air-pollution hazard when the discarded film is burned, as in an incinerator; alternatives to the salts should be considered. To improve health and safety inside the plant, it is likely that one of the most effective moves will be a wider use of robotics where working conditions are not suitable for humans.

Improved Performance. The purpose of materials science and engineering historically has been to improve performance by modifying existing materials and developing new ones. This activity will remain important. Demand will continue for higher-performance alloys, tougher glass and ceramics, stronger and tougher composites, greater magnetic strengths. But the task grows more complex as performance criteria come to embrace chemical and biological as well as mechanical and physical properties. Consumers and legislation, furthermore, are calling increasingly for materials and products that are more durable, more reliable, safer, and less toxic. To meet these requirements, a number of complex, materials-related phenomena must be elucidated. They include corrosion, flammability, thermal and photodegradation, creep and fatigue, electro-migration and electrochemical action, and biological behavior.

Functional Substitution. Functional substitution offers great opportunity in materials science and engineering. The aim is not simply to replace one material with a better one, but to find a whole new way to do a job. To join two metals, for example, one can develop not just stronger nuts and bolts, but adhesives. Jet engines replace piston engines and propellers in aircarft; telephones replace the mails for transmitting information. Functional substitution can lead to the revision of consumption patterns for materials and energy and, indeed, can inspire the creation of entirely new industries. Widespread use of nuclear or solar energy could yield enormous savings in the trans-portation of fossil fuels. The transistor started the solid-state electronics industry, which has led to technologies like computers, missile-control systems, and a broad range of industrial, medical, and leisure products. Challenging problems for functional substitution include: developing materials and techniques for new methods of generating and storing electrical energy; and finding functional substitutes and biological materials to replace human organs.

Product Design. The better we understand the properties of materials and how to control them, the more efficiently we can design them into products, providing materials and design specialists work closely together from the beginning of the design and development process. The resulting interplay may change apparent design restrictions radically and achieve more effective solutions to the design problem. Purposeful blending of materials and design expertise, moreover, can contribute significantly to conservation of materials. Appropriate knowledge sometimes allows safety margins to be narrowed without hazard, thus reducing the weight of material needed in the product. Where properties like strength and elastic modulus can be upgraded, the product can be made, sometimes, to contain significantly less material without corresponding loss in performance. An example is the use of textured steel sheet in automobile bodies. Design can also be improved as a result of clarifying the functional requirements of specific parts of a product. If only a surface must resist corrosion, for example, coating or cladding may cost less -- and may require less material -- than use of corrosion-resistant material throughout.

Recovery, Recycling. Facilitating the recovery and recycling of materials -- apart from new approaches to questions like collection and separation -- presents broad new problems in product design and materials selection. Product designs should ease dismantling and separation of components, but the rising costs of repair services tend to favor materials and products designed for replacement as whole units rather than for dismantling and repair. These conflicting pressures will have to be reconciled. Metals like those in a shredded automobile tend to be degraded with each recycle, although they may be quite suitable for applications less demanding than the original ones. The same is true of blended plastics, ceramics, composites, and glass. It is not clear that these problems can be solved without sacrificing performance. We must learn not only to recycle materials more efficiently; we must develop secondary and tertiary outlets for recycled materials whose properties no longer meet the requirements of primary functions. Extractive chemistry and metallurgy will be important in improving recycling processes, but better physical methods of separation are needed, too.

Materials Challenges in Areas of National Impact

Peoples and nations have many kinds of concerns and goals. Some aspirations, like "Life, liberty and the pursuit of happiness," are not truly reducible to tasks for science and technology. Others, like the desire for a strong domestic economy, clearly call for positive technical contributions. In connection with national concerns of the latter type, we have surveyed a number of specific challenges and priorities for materials science and engineering.

One of our approaches involved a novel questionnaire on Priorities in the Field of Materials Science and Engineering. The questionnaire was designed to obtain both qualitative and quantitative opinions concerning important problems and specialties in materials research. It was sent to nearly 3,000 professionals representing many disciplines and responsibilities. The 555 usable replies were analyzed

in detail to yield, for each of the nine Areas of Impact, a selection
of high-priority topics and specialties in applied and basic research
in materials. The methodology and illustrative results appear in
Appendix A. Among other questions, the respondents were asked to rate,
on a five-level scale, the overall importance of materials science
and engineering to progress in each of the nine Areas of Impact; some
of the results, calculated as explained in Appendix A, appear in
Table 13. No attempt was made to rank-order the relative importance
of the nine Areas of Impact to the nation.

The scope and nature of materials problems we have discerned
in seven of the nine Areas of Impact are highlighted in the following
brief descriptions.* Exhaustive treatments are beyond the intent of
this summary report, though considerable additional information was
obtained from the priority analysis (see Opportunities in Materials
Research, page 97).

Needs in Communications

Society continues to demand communications systems of greater
capacity, versatility, and reliability for many purposes: telephone;
radio and television program distribution; information processing,
storage, and retrieval; automatic billing, credit-checking, and other
operations of a cashless society; airline and hotel reservations;
police and fire departments; aircraft navigation and traffic control.

---

* These descriptions do not match precisely the Areas of Impact listed in
Table 13. Materials problems in Defense, for example, are documented
extensively elsewhere and so are not covered in the section on Space
(page 68).

TABLE 13

Relative Importance of Materials Science and Engineering
in Nine Areas of Impact

Area of Impact	Rating Number[*]
Communications, Computers, and Control	92
Defense and Space	92
Energy	90
Transportation Equipment	69
Health Services	69
Environmental Quality	68
Housing and Other Construction	67
Production Equipment	61
Consumer Goods	60

[*] Rating numbers were derived from analysis of replies to a priorities
questionnaire, using the methodology described in Appendix A. Note
that respondents were not asked to rate relative importance of the
areas themselves to national or other interests.

These demands on communication systems translate into requirements for transmissions at higher and higher frequencies and for hardware inherently more reliable than the existing terminal, switching, and transmission devices. Implicit in such needs is a variety of problems in electronic, optical, and related materials.

Existing communications technology relies heavily on electro-mechanical and reed-contact relays, wire and cable, and microwave radio transmission. These technologies operate under the traditional pressures to reduce costs while increasing reliability. Newer technologies rely heavily on solid-state electronics, particularly integrated circuits. Solid-state electronics, the basis of modern computer and communications systems, is inherently far more reliable and versatile than vacuum-tube devices. Progress in communications also requires the development of cheaper long-distance, broad-band transmission. Involved here will be further advances in microware transmission, including the use of satellites and waveguides, and optical communication technologies. Also necessary will be new switching methods that take advantage of the memory and logic capabilities of integrated circuits, magnetic-bubble and charge-coupled devices, and minicomputers. New customer services will call for cheap, reliable visual displays and data terminals to replace the more cumbersome cathode-ray tube and tele-typewriter. In this vein, the recently developed charge-coupled devices seem especially promising for solid-state television cameras and display systems.

In traditional equipment, there is steady progress toward economizing on materials through miniaturization and by developing

substitutes from cheaper materials. The development of textured spring
alloys, for example, has contributed to the miniaturization of electromagnetic relays; ferromagnetic alloys in miniature reed switches have
led to compatibility between the fast-switching capabilities of modern
electronics and the slower-switching operations of electromechanical
equipment; new plastics have been developed for wire insulation that
can withstand exposure to various working environments; and, in the
substitution area, the price of copper has stimulated the development
of aluminum wire and cable together with suitable connection technology.
An increasingly pressing problem is to find substitutes for palladium
in electrical contacts and for gold in thin-film circuits.

Solid-state electronics is a pace-setter in the materials field
for ultimate control over material quality, since electronic properties
are so sensitive to impurities and defects. The need for quality control is particularly severe in the manufacture of integrated circuits
and optical devices, especially as the dimensions of such devices become
smaller and smaller. Critical, too, are advances in the photolithographic-
polymer mask process used to lay down circuit pathways. The beams of
light employed for this purpose are very close to the maximum possible
definition, so that further substantial miniaturization of the circuits,
an important factor in cost reduction, is not feasible. For this reason,
the semiconductor industry is developing electron-beam lithography
processes, which promise ultimately to reduce the size of integrated
circuits from 100th to 1,000th the size of those possible with current
methods. Critical to the electron-beam approach is the synthesis of

improved mask and lithography materials and better understanding of the physical interaction between such materials and electron beams.

For transmission, especially in conduits under city streets, optical waveguides will permit great savings in space as contrasted to copper or aluminum wire and, at the same time, considerably greater channel capacity. Here, again, a critical need is close control of the composition and structure of the glass fibers involved. Optical communications systems, in addition, will require considerable materials work to develop reliable terminal device arrays. These arrays will comprise a hybrid technology -- integrated optics -- optical devices like light-emitting diodes, lasers, photodetectors, and optical waveguides driven or controlled by silicon integrated circuits.

Needs in Space

Steady improvement in the performance of materials is required to meet this country's current and projected priorities in space, which fall into three general areas: manned space flight, space science, and space applications. With the successful completion of the Apollo program, the manned space flight effort will emphasize the earth-orbiting Skylab and the earth-to-orbit Space Shuttle. The space science programs, likewise, will concentrate more on exploration of the earth's environment, the solar system, and beyond, using a variety of spacecraft and missions: the Orbiting Solar Observatory, the High Energy Astronomical Observatory, the Mariner missions to Venus and Mercury, the Pioneer missions to the vicinity of Jupiter, and the Viking Mars orbiter and lander. Space applications involve near-earth satellites for meteorology, communications, navigation, geodesy, and earth-resources surveys. Typical unmanned, earth-orbit spacecraft are the Nimbus weather satellite and the Earth Resources Technology Satellite.

An important requirement of the Space Shuttle is a thermal protection system that can be reused about 100 times. One aspect of this problem is that the craft's payload, and thus the cost of launching 1 pound of payload into orbit, is very sensitive to the weight of the thermal protection system. The leading candidate materials for the system are rigidized ceramic-fiber insulations protected with a ceramic coating. These materials are simple, light-weight, highly-efficient thermal insulators. They should survive repeatedly, without significant loss in performance, the maximum expected surface temperatures caused by aerodynamic heating during normal reentry. Should the ceramic insulations meet unexpected difficulties, however, other thermal protection materials must be developed as backups. These include superalloys, coated refractory metals, and ablators.

The Space Shuttle and other spacecraft will require materials that provide better mechanical properties per unit of weight. The same need applies to space payloads. These requirements call for greater use of alloys and composites having high ratios of strength to weight and of stiffness to weight. As the vehicles move deeper into space, they will require materials that resist radiation and extremes in temperature more effectively than do those now available. Improved surveillance and communications satellites demand optical materials of increased integrity, and mirrors that maintain their shape indefinitely (e.g., that have a near-zero coefficient of thermal expansion).

Research on lasers must be continued to realize the potential of such devices in communications. Materials with novel and superior electronic and optical properties are required for improved sensing devices or instruments. In particular, high-reliability sensors for earth-orbital spacecraft offer the benefits of rapid, continuous observation, greater freedom from weather disturbances, large-area scans for regional synthesis, better-quality data, reduced data-acquisition time, and lower cost. Finally, long-lived materials with extreme service capability are needed for advanced batteries and power-generation systems.

Needs in Electrical Energy

Electrical generating capacity in the United States has doubled every decade since about World War I. There seems little doubt that demand for power will continue to grow, and that measures to safeguard the environment and conserve fuels will raise the cost of power. This will intensify the mounting pressure to generate, transmit, store, use, and conserve electricity more efficiently, and it is here that materials science and engineering becomes essential.

Energy-related materials problems range in difficulty from those of coal-gasification processes, which are perhaps not crucial, to the uncertain but probably exceptional demands that will be imposed by thermonuclear fusion processes. Discussed here are some selected materials needs in superconductors for generators and transmission lines; in high-temperature gas turbines; in magnetohydrodynamic generators; in breeder reactors; in solar-energy conversion; and in materials and devices for storing electrical energy.

It is too soon to forecast the materials requirements of a fusion power system, which should offer cheap, clean, and practically unlimited energy. The scientific feasibility of the process has yet to be fully demonstrated and, if it is, engineering feasibility is unlikely to be shown before the end of this century. The fusion reaction would occur at hundreds of millions of degrees in a plasma confined by a magnetic field, no known material being up to the task. A gas-containment vessel and other components will be needed, however, and it must be assumed that the materials used will have to cope with intense heat and radiation, in particular, extremely high-energy neutrons.

Superconductors. Superconductors are materials that lose their electrical resistance below a characteristic and, thus far, very low transition temperature. The alloy with the highest known transition temperature found yet becomes superconducting at only 23.2° above absolute zero or about 250°C below freezing. Known superconductors must thus be cooled cryogenically using helium or hydrogen liquids.

Despite the costs and engineering problems of cryogenic cooling, superconducting alloys may allow the efficiency of conventional electrical generators to be increased and their size reduced. Size and efficiency depend in part on the intensity of the magnetic field that can be produced by a given volume and weight of electrical conductor. Superconductors like niobium-tin yield very high magnetic fields; progress is being made with this approach, but difficult fundamental and engineering obstacles remain. At some point, moreover, extensive high-power experiments will have to be done on full-scale generators, which are very expensive to build and evaluate.

Superconducting cable may ultimately reduce the cost of transmitting large amounts of power underground. The passage of direct current through superconductors generates essentially no heat, which would avoid the problems of dissipating heat in confined spaces. The hope is that, when transmitting very large blocks of power, the resulting savings would more than offset the cost of cryogenic cooling. Superconductors with higher transition temperatures than those now available are desirable. Even small increases in transition temperature can improve the economics of cryogenic cooling. Whether significant improvement is possible, however, cannot be predicted, since we still know too little of the fundamental relationships of the composition, structure, and properties of superconductors.

High-Temperature Turbines. Electric utilities have turned significantly to generating electricity with gas turbines, which can be obtained and brought on-line quickly, to make up shortages caused by delayed additions of nuclear and sometimes conventional generating capacity. The thermal efficiency of the turbines, however, is only 25 to 30 percent when operating on natural gas at about 870°C. If the temperature can be raised to between 1,100° and 1,300°C, an overall efficiency of 40 to 45 percent is predicted to be possible. The key to achieving such efficiency will be new or improved materials for the turbine parts, heat exchangers, and related apparatus. These materials must be strong and readily fabricated and, at the required temperature, must resist oxidation, impact, and thermal fatigue. It would be advantageous, too, if they could withstand the abrasive gas produced by burning pulverized coal, thus making the latter technology more attractive. Such conditions cannot be met even by the most sophisticated alloys developed for turbine blading, so that ceramic materials will be necessary. Those being studied include silicon nitride, silicon carbide, and composites of silicon carbide or aluminum oxide fibers in a metal matrix. Selection of the optimum materials must await more comprehensive knowledge of their properties and performance.

MHD Generator. In the magnetohydrodynamic (MHD) generator a hot conducting gas replaces the rotating generator of conventional equipment. In either case, a magnetic field induces current in the conductor moving through the field. In the MHD device, current generated in the gas by passing it through a transverse magnetic field is drawn off by electrodes inserted into the stream. An MHD generator should achieve a thermal efficiency of 50 to 60 percent, but the materials problems are unusually severe. Current thinking envisions zirconium oxide for the electrodes, and aluminum, zirconium, and magnesium oxide for the burners. The conductive gas could be combustion gases from coal, seeded with potassium sulfate and maintained at 2,000° to 2,400°C. We know far too little to be able to predict the behavior of this combination of materials at the temperatures involved. Reliable design criteria and materials selection will come only through coordinated materials research and engineering in several areas.

Breeder Reactor. Development of commercial breeder reactors, which may be the best hope for a relatively nonpolluting, long-term power supply, is a major interdisciplinary task. The fuel elements alone present formidable materials problems (and coping with radioactive wastes is not the least of many other difficulties). Probably they will resemble today's power-reactor fuel elements, which are thin-walled tubes, about 4 meters long, filled with uranium dioxide pellets. The breeder reactor, however, probably will use uranium-plutonium oxide pellets as fuel.

Effective fuel-element design requires that a number of characteristics of the pellets be clarified. These include elastic and plastic deformation, fracture, corrosion, effects of change in system composition, diffusion, and crystal defects. Such characteristics will be modified in complex ways, moreover, by radiation damage and large temperature gradients. Obtaining the needed understanding of the properties of this new fuel will occupy many scientists and engineers for several years, but the work is essential if breeder reactors are to be operating commercially by 1990, the current target in this country.

Solar Energy. Solar energy is chemically and thermally nonpolluting, and a number of approaches to harnessing it are being explored. These include conversion to electricity, directly or by steam turbine or other types of generators; heating and cooling homes and providing domestic hot water; and extracting energy from sun-powered phenomena such as winds and thermal gradients in the oceans. Progress in materials is important to such schemes. To convert sunlight to electricity, for example, requires large arrays of efficient, low-cost solar cells or collectors. Existing materials cannot fully meet the requirements.

The present silicon solar cells are effective in the space program, but widespread conversion of sunlight to direct current requires

more practical devices. The highest efficiency achieved yet is 18 percent -- 4 to 5 percent above the best previous level -- in a gallium
arsenide cell; also under study is cadmium sulfide. Silicon, whose
conversion efficiency is about 11 percent, remains the only cell material
available commercially. The efficiency of solar collectors can be
raised by coatings that absorb the sun's energy much more readily than
they re-emit it. Collector coatings may have to function at perhaps
500°C, however, to improve the thermal efficiency of the system, and
existing coatings may not be sufficiently stable. Molten salts, which
are candidates for storage media, will cause corrosion problems in
piping and containment vessels. Another scheme is to use wind-
generated electricity to electrolyze water to hydrogen and oxygen,
which could be stored and recombined at will in an energy-producing
fuel cell. Materials problems here include fuel-cell catalysts and
hydrogen embrittlement in piping. The latter problem could be serious
in the proposed "hydrogen economy," in which hydrogen would be a major
fuel.

Energy Storage. Environmental concerns have prompted new efforts to
develop inexpensive, reliable, nonpolluting means of converting and
storing energy. Among the possibilities are improved electrochemical
devices: fuel cells, primary cells (flashlight battery), and secondary
cells (storage battery). Each of these converts stored chemical energy
to electrical energy upon demand. The secondary cell, when recharged,
converts electrical energy to chemical energy, stores it, and reconverts
it to electrical energy upon demand.

Electrochemical devices potentially could replace the internal combustion engine in automobiles. Research in this area has produced promising secondary cells that include the sodium-sulfur battery announced several years ago. The anode in this device is liquid sodium; the cathode is liquid sulfur (in contact with carbon felt). The main innovation in the battery is the solid-state electrolyte, beta-alumina. It allows unusually high transport of sodium ions, which diffuse through it to produce electric current when the battery is connected to an external load. The cell is recharged by reversing the process.

The sodium-sulfur battery, though a significant development, is not well adapted to vehicle propulsion. Its operating temperature is high, around 300°C, and metallic sodium would be hazardous in a crash. Many other materials are being studied for use as electrodes and electrolytes in electrochemical devices. The beta-alumina, for example, only suggests a crystallographic structure that should be explored for other solid-state electrolytes. The ultimate material, if there is one, probably will be found in quite different classes of structures.

Needs in Transportation

Advances in transportation in this country are hampered primarily by social, political, and economic factors. But while materials are not limiting, they do present important opportunities and problems. Transportation systems consume enormous amounts of metals, concrete, plastics, and other commodities; in the federal highway system, maintenance costs for bridges alone are estimated at $6 billion over the next decade. In such a context any technical advance can affect costs markedly. More specifically, the automobile industry faces immediate materials difficulties induced not only by rising costs and competition, but also by safety and emission standards. Improved materials, in addition, would facilitate progress in mass-transit systems, marine transportation, and aircraft.

Automobiles. The sheer size of the automobile market makes its materials needs critical. A weight increase of 1 pound per vehicle adds 5000 tons annually to the industry's consumption of materials. Automotive transportation accounts for about 45 percent of all United States transportation expenditures and 84 percent of total passenger-transportation costs. The passenger car is used for 82 percent of commuting to work, and city streets are estimated to handle more than half of the vehicle miles traveled nationally. Pressure is rising to reduce the reliance on the automobile, particularly in cities, but the change seems unlikely to come rapidly.

In few industries is competition among materials as fierce as in the automobile industry, which converts some $5 billion in materials annually into 10 million vehicles and will struggle to lop pennies from the manufacturing cost per car. The competition is being intensified by vehicle safety and emission standards. New equipment mandated into autos in 1971-76 includes bumper systems, safety devices, emission controls, and heavier suspensions and body components to support the extra weight. These changes could increase the weight of a typical passenger car by 10 percent or more. Greater weight requires more power and thus control of a higher level of emissions. Reduction in weight, consequently, has become a major design problem, along with safety and emission standards, and each of these problems is materials-intensive.

Meeting safety standards will require additional structural members, which is prompting research on more effective use of high-strength steels and on the light metals, aluminum and magnesium. For a high-integrity safety car, steel may be too heavy to use in the large number of structural members required for good energy absorption. Aluminum would be the natural successor, although steel honeycomb could be competitive if its fabrication costs were reduced.

High-strength, low-alloy (HSLA) steels are prime candidate materials for reinforcing ultrasafe passenger compartments. A return to the separate reinforced frame will require more steel, some of it probably HSLA. More steel will be needed also in door safety beams, double roofs, and bumper systems. With proper design, however, new high-strength steels could help to reduce weight. If a part now made of carbon steel sheet can be made 15% thinner, for example, it can usually be made of currently-available HSLA steel -- to have the same functional strength at the same cost but with less weight.

Among the emission-control problems is the need to find durable materials that will last for the 50,000 miles required for auto-exhaust systems. The only proven materials for this use, muffler stainless steel, contains about 11 percent chromium, of which the United States has no domestic supply. The consequent increased demand for chromium in exhaust systems may require materials substitutions in other uses.

Severe difficulties exist with catalytic emission-control systems, as indicated by the one-year delay granted recently by the Environmental Protection Agency for meeting the 1975 standards. Catalysis, as mentioned earlier, seems not to be the best long-term solution to the emissions problem, but the possibility of at least interim use has stimulated a considerable effort on catalytic materials. Platinum, the best material available now, would have to be largely imported, and spent auto catalyst probably would have to be regenerated and recycled. Other catalysts tested thus far are less effective than platinum and may be simply inadequate.

To hold down costs the automobile industry, among other measures, is working on more efficient processes, such as high-speed induction sintering of powdered metals, continuous casting or electro-forming of metal sheet and strip, and reuse of plant scrap in manufactured products. Research is likely on more quantitative means of evaluating the acceptability of materials during production -- as in the melt, ingot, or powder stage -- to reduce the processing of rejectable materials. Expanded research is in prospect on joining, particularly the use of adhesive bonding and inertial and electron-beam welding in large-volume production. Already well established is the effort to reduce materials consumption by sharper design and improved functional characteristics.

Mass Transit. The main structural requirements of mass-transit systems can be met, in general, by state-of-the-art materials, although stronger, more durable light-weight structures are important for transit cars as well as for autos. Many improvements are needed in specialized subsystems: electrical controls, storage-battery components, electro-deposited finishes, and nonsmoking, self-extinguishing polymers. Lower-cost electrical insulations are desired, for example; present resin systems require considerable tailoring to achieve the best combination of dielectric strength, life, flammability, and smoke behavior.

An important area is the development of solid-state power components. As the speeds and power of rail vehicles increase, the trend in propulsion will be toward greater electrification and alternating-current traction motors. The related component problems lie in three areas:

- Energy transfer -- as from the power system grid to the moving vehicle and in reverse for braking, the latter being particularly significant for cooler subways.

- Conversion of power on board the vehicle for various purposes -- power either transferred from the grid or generated on board.

- Propulsion units -- either alternating current rotation motors for driving wheeled vehicles or linear motors for driving air-cushioned or magnetically levitated vehicles.

Marine Transportation. Among important materials problems in marine transportation is corrosion resistance in propulsion systems. The relative availabilities of various fuels are somewhat uncertain in the future, but it must be assumed that the cheapest residual oils will be used in marine machinery: boilers for steam turbines, marine diesels, and gas turbines. The materials problems involve hot corrosion in the presence of salt-containing air, sulfur, and some components of ash, like sodium and vanadium. Ash deposition is also harmful. These difficulties are being explored by the U. S. Maritime Administration in cooperation with marine equipment suppliers. Further in the future will come materials bottlenecks in nuclear propulsion. Current work on both land-based and sea-going reactors should provide at least partial solutions in this area.

Aircraft. In airborne transportation, significant progress can be made by improving the thrust-to-weight ratio of jet engines. The materials with the best potential for achieving this goal are composites. Carbon- or boron-reinforced polymers are the most likely composites for use in low-temperature fans. These materials are quite promising for reducing engine weight. Current research goals are tough, readily processible polymers that can withstand 250°C. Polymers may be required later for service at 300° to 500°C, the temperatures that will be encountered in higher-speed, multistage fans. An important backup to the graphite-polymer composites is aluminum reinforced with boron fibers.

Composites for use in the turbine section of a jet engine present still greater difficulties. These materials must be able to operate for thousands of hours at more than 1,000°C; they must be light, strong, and dimensionally stable and must have excellent oxidation resistance. One possibility is "natural composites" or oriented eutectics, in which the reinforcing and matrix phases are produced automatically in certain favorable cases by carefully controlling the conditions of solidification. These composites promise major advances in high-temperature metals.

Such systems include nickel-base alloys reinforced with tantalum carbide or nickel-niobium (Ni₃Nb) and cobalt-base alloys reinforced with tantalum carbide. Turbine materials for operation above 1,300°C probably will have to be made of coated refractory metals.

Needs in Health Services

Only recently have medical and materials people begun to work seriously together to make or modify materials specifically for medical use. The soft, water-swellable gel, Hydron, resulted from collaboration among polymer chemists and surgeons; ultrapure, medically acceptable silicone rubber came as an industrial response to a medical need. Both materials are for surgical implants. A longstanding effort on bio materials is the dental-research program of the National Bureau of Standards and the American Dental Association. This interdisciplinary effort has produced improved dental materials, standards, test methods, and instrumentation. Although other such cases exist, the general tendency is still to rely on materials developed first for nonmedical purposes.

Work on biomaterials is not extensive yet, but in the coming decade may become an exciting thrust of materials science and engineering. Good progress, however, will require that several obstacles be overcome:

- Industrial interest is inhibited by relatively small markets and by shifting national goals in health care.
- Lack of basic knowledge forces us to rely mainly on empirical solutions to important biomaterials problems.

- The unusual interdisciplinarity of the field requires new perspectives, particularly in the university sector.

- Standards and testing capabilities for biomaterials are absent or inadequate.

Small Markets. The markets for surgical and dental supplies in this country in 1970 were each in the range of $250 million to $300 million. These products -- e.g., sutures, syringes, and dental cements -- are made by aggressive companies with active though highly proprietary development programs. Research on specialized biomaterials, on the other hand, is restricted because of the smaller markets in cardio-vascular equipment ($127 million in 1970) and in catheters, artificial organs, and heart-lung equipment (each between $20 million and $25 million). Still, workers in these smaller areas have achieved a degree of success by innovative use of materials like silicone rubber, Dacron, and polyvinyl chloride in heart valves, oxygenators, and vascular (circulatory system) replacements. Industrial titanium, tantalum, vitallium (a cobalt-based alloy), and stainless steel are also used quite commonly in prosthetic and orthopedic devices.

Research in specialized biomaterials is hampered, moreover, by its decline in relative national priority during the past few years. Higher priority now is assigned to improvement in general health-care services. Progress in materials is most likely to contribute to the achievement of this goal when it is translated into markets for medical-service equipment, such as diagnostic and monitoring devices based on fundamental knowledge of the physical and chemical properties of biological substances.

Gaps in Knowledge. A notable gap in basic knowledge is clear in the artificial kidney. After 30 years of steady improvement, development of the device has reached a plateau. Further development is stalled because we know too little of the diffusional and surface properties of polymers to build better membranes and blood-access connectors. The problem has become even more pressing since July 1, 1973, when Medicare began to pay 80 percent of the cost of kidney dialysis after the first three months of treatment.

There are major unknowns in the reactions of blood with surgical implant materials. Nearly all nonliving and many living materials are thrombogenic: they induce blood-cell damage, clot formation, and protein destruction. Studies of surface charge, surface free energy, and similar properties of various materials have turned up few if any correlations. Surfaces containing bonded heparin (an anticoagulant) and aqueous gel surfaces are compatible with blood to a degree, but have yet to be proven clinically. The interdisciplinary approach of materials science and engineering clearly could create a major impact in this area.

New Perspectives. The range of disciplines needed to solve problems in biomaterials calls for new perspectives at the university level. Scientists have shown that, for example, bone will grow comparatively deeply into pores in the surfaces of metals and ceramics when pore diameter is 100 microns (0.004 inch) or more. There is some hope that implants of suitable surface porosity can be firmly and stably grafted to bone in this way. The work, however, requires experts in ceramics, powder metallurgy, orthopedics, and biomechanics. Such a variety of disciplines means that the team and the funding must be relatively large to ensure significant progress. A situation of this kind is a marked departure for those funding agencies and university administrators who are more accustomed to dealing with academic research focus in units of one faculty member.

Lack of Standards. A major difficulty in biomedical materials is the lack or inadequacy of agreed-upon standards and test methods. The problem is likely to persist so long as evaluation of many of these materials depends on the surgical profession, since the system is not designed to obtain the comprehensive life-test data needed for surgical implant materials. Federal agencies are beginning to encourage their contractors and grantees in biomaterials research and development to adopt the methods of industrial materials science and engineering, including scientific standards of measurement. Lack of basic knowledge, however, often leads to less than adequate test methods. The bio-materials field could be significantly advanced by the establishment of a suitable research center concerned with the evaluation of bio-materials, together with the development of associated test methods and standards.

Despite these growing pains, some activity is under way in bio-materials. Many materials have been screened in the search for non-thrombogenic vascular replacements. The leading candidate materials, thus far, are composite structures of non-thrombogenic hydrogel, carbon, or cell-seeded microfiber surfaces on silicone rubber substrates, and some segmented polyurethanes. Plastics reinforced with glass fibers are being studied for orthopedic service. A variety of membrane-fabrication techniques are being used, mainly with common materials, to improve the performance of economical oxygenator systems. And a new, still highly experimental use of biomaterials is in biologically active surgical implants. Enzymes bound to the surface of such an implant, for example, may catalyze specific biochemical reactions in the body.

Needs in Environmental Management

Many of the nation's environmental problems reflect the customary inattention to the materials cycle. Industry, including the materials community, has tended, understandably, to optimize only that segment of the cycle that deals with the incoming material through to the outgoing product, especially the parts of that segment where optimization will reduce costs. This approach creates environmental impacts that by now are well recognized. Few would question the need to broaden the span of materials management to the full cycle, from primary resources through disposal and reclamation, including the side effects such as emissions to the environment. Materials science and engineering can do much to ease the transition.

Manufacturing. Knowledge of properties of materials can contribute to the development of manufacturing properties that generate less waste but preserve the functional properties of the materials involved. A general approach mentioned earlier is to shape products directly from fluid (liquid or powder) materials instead of from solids. A more specific example is the printed circuit board, which the electronics industry makes from a board covered completely by a copper film. The film is etched away chemically to leave copper only along the desired circuit pathways. The process produces, industrywide, more than 5 million pounds per year of dissolved copper in wastes. Most of the metal is recovered, both because it is costly and because otherwise the wastes would present a severe disposal problem. New processes have been devised, however, in which copper is deposited in the first place only along the desired circuit pathways. The circuit pattern is first printed on the board with a special ink or with a polymer coating that can be optically activated and chemically developed in the shape of the pattern. The copper, deposited from a chemical bath, adheres only to the pattern delineated on the board. Both the ink and polymer processes avoid the waste problem and also reduce the costs of starting materials.

Recycling. Materials scientists and engineers can create materials, or combinations of materials, that function as required and, at the same time, are amenable to product designs that facilitate recycling. An everyday instance is the glue used in some types of cardboard boxes. This adhesive gums up the paper-processing step so that the paper in the boxes cannot be recycled economically. Careful study of precisely how glue effects a bond, in terms of its composition and structure, almost certainly would yield an adhesive more compatible with reprocessing. Metals provide a second example. Some alloying substances and coatings degrade the base metal to the point where it cannot be recovered economically from the alloy. Similar problems exist with polymer blends. Materials research should create alternative materials systems that fill the design function at competitive cost and still permit economic reclamation

Waste Conversion. Wastes can often be converted to useful materials, but the requirements of environmental protection and the related skewing of the traditional economic framework call for more concentrated effort. Examples of potential uses of wastes include the manufacture of brick from fly-ash or coal-ash slag combined with suitable binders. One new method of making brick combines almost any solid inorganic material with a small amount of portland cement and a proprietary chemical accelerator. The mixture is molded at high pressure to give a brick whose properties adapt it to new construction techniques that reduce labor costs.

A waste of unusual potential is lignin, the cellulose-bonding material in wood. Lignin and other soluble components amount to some 55 percent of the weight of the tree. The lignin, about 25 million tons of it annually, is discarded with waste pulping liquor by paper producers. A small amount of lignin is recovered as lignosulfonates, which are used by industry as surface-active agents. An attractive possibility, however, is to employ lignin as a bonding agent in wood products, where it would function as it does in nature.

The ability to use lignin in this way would represent a marked gain, both in conservation of resources and in environmental improvement, but the problem is particularly difficult. It has not responded to concerted scientific attack for some years. New hope lies currently in the use of sophisticated methods, such as high-voltage electron microscopy, to probe the chemical structure of lignin as it exists in the tree. Scientists to date have had to study the compound largely after it emerged from the severe chemical attack of the pulping process.

Organic wastes might provide a raw-material base for biodegradable plastics, although the costs involved are quite unfavorable today contrasted with plastics made from petroleum and natural-gas liquids. Newsprint and waste cotton fabric might be recycled to become a raw-material base for cellulosic plastics. Ethylene, the leading monomer for plastics today, might be manufactured by fermenting sugars or hydrolyzed cellulose to ethyl alcohol, which could be cracked to ethylene.

Packaging. Noteworthy among environmental concerns are packaging materials. Consider the simple beer can, a common element of litter. For economic reasons, producers have shifted from iron-based cans, which rust away eventually, to aluminum cans, which last almost indefinitely. A container material that degrades quickly and naturally could solve the litter problem. It might be costly, on the other hand, and be made from a relatively limited resource like petroleum. Iron or aluminum can be reclaimed without great difficulty, if at some cost, and reclamation is perhaps the pragmatic course at the moment. In any event, the packaging problem illustrates an environmental role of materials science and engineering applied to the materials cycle.

Needs in Housing

The materials-related problems in housing, even those of broad import, hamper progress much less than do legal, economic, and cultural constraints. Easing these constraints was one of the goals of Operation Breakthrough, sponsored by the Department of Housing and Urban Development (HUD). An official of HUD has said of Breakthrough,

"...very little of what we are doing involves basic research or totally new hardware technology." HUD hoped to "...facilitate the...use of technology which currently exists but has not found extensive use by the housing industry."

Predicting Behavior. A major bar to the introduction of new materials in housing is the unreliability of short-term tests for predicting long-term behavior such as weathering. A related obstacle is the lack of data on the performance of materials and components in service. Information is scattered and uncorrelated; a sound compilation is not available to the housing industry. Such data are basic, not only to the use of current materials but for developing short-term tests and otherwise predicting the behavior of new materials.

A second general need is a more fundamental basis for predicting the effects of fires in buildings. Laboratory tests have been devised to measure such parameters as speed of ignition, rate of flame spread, smoke evolution, and penetration of fire through walls and partitions, but the validity of these empirical tests is too uncertain. Their frequent appearance in building codes is due largely to the lack of anything better. The peculiar needs of flammability research could be well served by a center devoted to this subject.

Performance Criteria. The estimated 4,000-6,000 building codes in this country impede progress partly because they are out of date and more importantly because they differ among themselves. The resulting effects are to limit both the range of usable materials and the size of their markets; both factors restrict the effort likely to be expended

on new materials. The probable solution is to replace building codes
with performance criteria. Such criteria would specify degrees of
safety, durability, livability, and the like, but would leave design
and materials selection to the engineer. (The National Bureau of
Standards developed "guide criteria" in the course of its evaluation
of the 22 housing prototypes created by industry under Operation
Breakthrough.)

Replacement of building codes by performance criteria is
expected eventually to permit the development of modular home-building
systems. These could consist of standard components, assembled offsite
into a variety of configurations. Offsite manufacture will reduce
costs, but perhaps by no more than 20 percent of the cost of the home,
including land, financing, and so forth. Performance criteria and
testing, however, will help resolve other problems: protecting the
public against poorly designed and unsafe structures; establishing
standards of construction suitable for Federal Housing Administration
mortgages; and developing more livable communities of homes.

Thermal Insulation. The National Bureau of Standards has estimated
that energy consumed in residential heating and air conditioning could
be reduced, nationwide, by as much as 50 percent from current levels by
better insulation and construction. The technology is available, and
insulation is being upgraded.

Offsite Assembly. Developments in housing materials will come most
likely in response to demands of industrialized offsite assembly processes.
A number of laboratories, for example, are seeking materials capable of

joining housing components satisfactorily for long-term service. Other problems include the fire-retardation properties of synthetics, which generally will have to be less expensive before the materials can enjoy widespread adoption in housing. Materials such as gypsum board, plywood, concrete, glass, and aluminum, on the other hand, are remarkably inexpensive and unlikely to be displaced soon in their traditional functions.

The persistence of familiar housing materials is evident in the mobile home industry. Mobile homes are made in sufficient volume -- the number sold in 1972 amounted to 20-25 percent of the housing starts that year -- to be well adapted to industrialized assembly. They are not subject to building codes, moreover, and thus offer unusual flexibility in materials selection and assembly techniques. Even so, the industry has not seen the emergence of spectacular new families of materials. Aluminum sheet, hardboard, and the like are in common use. More advanced concepts are found perhaps in bathroom materials, where the traditional cast iron and porcelain have given way largely to reinforced plastics. These are mainly glass-fiber reinforced polyesters with gel coats. They are light and strong and withstand the handling involved in assembly and transportation. On the other hand, they scratch more readily than porcelain, and are subject to cigarette stains and charring although they meet fire safety requirements.

New building materials have appeared in recent years, of course, for applications other than in mobile homes. Prominent among them are polymeric products such as vinyl tile, polymer-paper laminates for working surfaces, and sealants and gaskets for installing large sheets of glass and prefabricated components. A product specifically of materials science and engineering, polymer latex-modified portland cement, has experienced modest but growing success in the decade it has been on the market, despite its relatively high cost. More recently, a high-strength concrete reinforced with steel or glass fibers has been developed. These newer materials have succeeded as a rule by producing economies, as in assembly or maintenance, that offset their higher initial costs. And since the greatest economies in housing probably will stem from offsite assembly, novel materials seem likely to be tied to that method of construction.

Needs in Consumer Goods, Production Equipment, Automation

In addition to the preceding illustrative studies of materials in seven areas of impact, we have analyzed the qualitative responses to the COSMAT priorities questionnaire in terms of specific research needs in all nine of the areas surveyed. In consumer goods and production equipment, the two areas not covered above, the responses mentioned certain requirements consistently. These include, for consumer goods, greater durability (both physical and chemical), less flammability, and greater safety, reliability, serviceability, and maintainability. A clear need exists also for better tests for these characteristics. Materials problems mentioned consistently for production equipment include longer-lasting, higher-speed machining devices, both metallic and ceramic (e.g., grinding wheels), better joining methods, and greater high-temperature strength.

We have noted earlier the attractive opportunities in a special area of production equipment: automation and robotics. These opportunities exist not only in production and manufacturing, but also throughout the service areas of the economy -- mail sorting, billing, typesetting, weather forecasting, health checkups, traffic control. Automation techniques in all of these areas include a common approach: the generation and processing of information to provide or display data in useful forms or to control servomechanisms . Myriad opportunities can be discerned in primary information-generating devices or sensors, which will depend on the nature of the quantity or physical property to be measured, the object to be sensed, or the pattern to be diagnosed. Nearly always these sensing techniques must be nondestructive. They must rely, therefore, on the effects of the interaction of matter with various kinds of radiation -- optical, electromagnetic, ultrasonic, and others. Progress in this field clearly will require the most sophisticated knowledge of materials and of spectroscopy in its broadest sense.

The signals generated by the primary sensing device usually must be processed, analyzed, and correlated by a computer or, increasingly, a mini-computer, itself a product of sophisticated materials science and engineering in its integral circuits and memory devices. Once in useful form, the information can be printed out, visually displayed, or used to control a machine or servomechanism. Opportunities for improvement lie both in visual displays and in computer-controlled machines. The latter can range from simple mechanical transducers -- to control a valve, for example -- to complex robots that can simulate some of the routine actions of human beings.

The development of this type of automation will require new devices, particularly optoelectronic ones, and solid-state electronic circuits with associative memory and learning capability for parallel processing. Especially promising avenues for further research appear to be semiconductor lasers and light-emitting diodes, magnetic-bubble devices, charge-coupled devices, reversible photosensitive materials, liquid crystals, optical modulators and deflectors, and various functional components such as amplifiers, timing circuits, and shift registers. Advances in servomechanism design will call for the combined talents of electrical and mechanical engineers, but often these devices and machines will also place stringent demands on the material of which they are made, especially when the equipment must work relia for long periods in hostile environments.

Automation is a very broad interdisciplinary area and is likel to become more so. It combines the knowledge and skills of materials scientists and engineers with those of the information community -- mathematicians and statisticians, as well as computer hardware and software engineers. The economic and social implications of switchin to automation in a given operation, moreover, can call also for the expertise of economists and social scientists.

Goal-Oriented Materials Research
Bearing on Areas of National Impact

COSMAT analyzed several thousand write-in comments from materi professionals to derive a list of goal-oriented research topics that high priority in the nine national areas of impact. The topics selec included various properties of materials, classes of materials, proce and applications (Table 14).

TABLE 14

Goal-Oriented Materials Research
Bearing on Areas of National Impact

Analysis of several thousand write-in comments from materials professionals
indicates that the topics below rate high priority in research in the areas of
impact shown. Where applications are listed, the meaning, generally, is that
new materials and processes are needed to advance the application.

<u>Communications, Computers, and Control</u>

Memories; visual displays, semiconductors, thin films; integrated circuit processes, yields in large-scale integration, component reliability; optical communication systems; defect properties of crystals; chemical and surface properties
of electronic materials; purification; crystal growth and epitaxy; joining
techniques; contacts; high-temperature semiconductors.

<u>Consumer Goods</u>

Durability; visual displays; corrosion; mechanical properties; improved strength-
to-weight packaging; recyclable containers; high-strength glass; plastics;
plastic processing; composites.

<u>Defense and Space</u>

Mechanical properties; lasers and optical devices; energy sources; heat resistance; corrosion; radiation-damage-resistant electronics; composites; turbine
blades; heat shields; thermal-control coatings; nondestructive testing; higher
strength-to-weight-ratio materials; reliability; materials for deep-sea
vehicles; joining.

<u>Energy</u>

Battery electrodes; solid-state electrolytes; seals; superconductors; electrical
insulators; mechanical properties; radiation damage; high-temperature materials;
corrosion; joining; nondestructive testing.

<u>Environmental Quality</u>

Less-polluting materials processes; pollution standards; recyclability; reduced
safety and health hazards; extraction processes; catalysts; secondary uses for
discarded materials; sorting processes; nondestructive testing; noise reduction.

<u>Health Services</u>

Implant materials; membranes; biocompatibility; medical sensors; material
degradation.

<u>Housing and Other Construction</u>

Prefabrication techniques; corrosion; cement and concrete; weatherability;
flammability.

<u>Production Equipment</u>

Friction and wear; corrosion; sensors; automation.

<u>Transportation Equipment</u>

Corrosion; pollution control; high strength-to-weight ratios; high-strength,
high-temperature materials; impact resistance; catalysts; adhesives; superconductors; lubricants.

OPPORTUNITIES IN MATERIALS RESEARCH

Priority Analysis

In order to gather many viewpoints on opportunities in materials research, both basic and applied, COSMAT solicited the opinions of a broad cross section of the technical community both on specific materials topics that deserve attention and on relative priorities for research among various classes of materials, materials properties, and processes. These inquiries went to the presidents of materials and materials-related technical societies, to pertinent Gordon Research Conferences, and to individuals known for their work in the materials field. In all, information was received from nearly 1,000 persons, including the 555 usable responses to the COSMAT questionnaire on Priorities in the Field of Materials Science and Engineering. The information was handled by two task forces: one analyzed the quantitative responses to the priority questionnaire; the other developed brief descriptive summaries of some of the research opportunities that were identified most frequently. The main results of the quantitative analysis appear below, followed by the descriptive summaries.

The methodology used in analyzing the priority questionnaire is described in Appendix A.

To assess priorities in applied research, each respondent was asked to indicate on a scale of 1 (very high) to 5 (very low) the priority that should be given to applied research and engineering in a given materials specialty (out of a list of 46) to assure progress toward a national objective (nine Areas of Impact and 52 Subareas) in which he claimed to be knowledgeable. Other information from the respondent allowed the degree of his familiarity in each instance to be taken into account in the analysis. The results of this study, integrated for each of the nine areas of impact, are summarized in Table 15. (See Appendix A for more information on priorities for applied research in various areas of impact and lists of specific research topics identified as having high priority.)

It is evident that some materials specialties are considered to be a high priority in certain areas of impact, but not in others. A few specialties, on the other hand, appear to have very broad relevance (Table 16). We would emphasize that the overall priority for a specialty cannot be established simply by totaling the stars across Table 15; this would presuppose that all areas of impact have equal levels of material priority and correspond to sectors of comparable importance to the nation's well-being. Rather, given the goal of advancing a _selected_ area of impact, Table 15 indicates the relative priorities of the materials specialties in that context.

Respondents were asked also, in connection with each materials specialty, to assign priorities to _basic_ research problems not necessarily identified with any particular area of impact. The results differed somewhat in emphasis from those for applied research, but the two overlapped considerably; problems described under the basic research heading were often the same as those described by others under the applied research heading.

The questionnaire results for priorities in basic research are summarized in Table 17. The various types of ratings designated in the right-hand columns are described in Appendix A. The materials specialties

TABLE 15

Priorities for Applied Research in Materials by Area of Impact

(x -- indicates above-average priority;
xxx -- indicates highest priority.)

	Communications, Computers, Control	Consumer Goods	Defense & Space	Energy	Environmental Quality	Health Services	Housing & Other Construction	Production Equip.	Transportation Equipment
PROPERTIES									
Atomic Structure (Crystallography and Defects)	xx		xx	xx		x			xx
Microstructure (Electron Microscope Level)	xx	x	xx	xx		xx		x	x
Microstructure (Optical Microscope Level)	x	x	x	x		x			
Thermodynamic (Phase Equilibria; Change of State, etc.)	x	x	x	xx			xx	x	xxx
Thermal (Thermal Cond., Phonons, Diffusion, etc.)		x	x	x					
Mechanical and Acoustic (Strength, Creep, Fatigue, Damping, etc.)	xx	x	xxx	xx		xx	x	x	x
Optical (Emission, Absorption, Luminescence, Excitation, etc.)	xx	x	x	x		x			
Electrical (Cond., Electron Trans., Ionic Cond., Thermoelec., Injection, Carrier Phen.)	xxx		x	x					
Magnetic (Ferromagnetic, Resonance, Paramagnetic)	x	x							
Dielectric (Ferroelectric, Breakdown, Loss, Piezoelectric, etc.)	xx	x		x					
Nuclear* (Radiation Damage, Absorption, Surface States, Catalysis, etc.)	x		x	x					
Chemical & Electrochemical‡ (Corrosion, Battery Phen., Oxidation, Flammability, etc.)		x	x	xxx	xx	xxx	xx	x	xx
Biological (Toxicity, Biodegradibility, etc.)	x	x	x		xx	xxx	x	x	x
MATERIALS									
Ceramics	xx		x	xx	x	x	x		
Glasses and Amorphous Materials	xx	x	x		x	x	x		
Elemental and Compound Semiconductors	xxx		x	xx					x
Inorganic, Non-Metallic Elements and Compounds	xx		x	x	x		x	x	
Ferrous Metals and Alloys			xx	x	x	x	x	xx	xx
Non-Ferrous Structural Metals and Alloys			xx	x	x		x	xx	xx
Non-Ferrous Conducting Metals and Alloys	x		x	x					
Plastics	x	xxx	xx	x	xx	xxx	xx	x	xx
Fibers and Textiles		x		x	x	xx	x		x
Rubbers		x				xx			xx

TABLE 15

Priorities for Applied Research in Materials by Area of Impact (con't.)

	Transportation Equipment	Production Equip.	Housing & Other Construction	Health Services	Environmental Quality	Energy	Defense & Space	Consumer Goods	Communications, Computers, Control
Composites	xx		xx	xx	x	x	xx	x	x
Organic and Organo-Metallic Compounds	x	x	x	xx	x	x		x	xxx
Thin Films	xxx	x		x				xx	x
Adhesives, Coatings, Finishes, Seals	xxx	x	xx	x			xx	xx	x
Lubricants, Oils, Solvents, Cleansers	xxx	xx					x		
Prosthetic and Medical Materials			x	xxx					
Plain and Reinforced Concrete			x		x			xx	
Asphaltic and Bituminous Materials			x	xx	x				
Wood and Paper									
PROCESSES									
Extraction, Purification, Refining					xx	x	x	xx	xx
Synthesis and Polymerization				xx	x	x	x	xx	xx
Solidification and Crystal Growth	x		x			x	x	x	xx
Metal Deformation and Processing	xx	x	xx	xx		x	x	xxx	x
Plastics Extrusion and Molding	x	x				x	x	x	
Heat Treatment	xx	x	xx				x	x	x
Material Removal (Machining, Electrochemical, Grinding, etc.)	xx			x		xx	xx	x	x
Joining (Welding, Soldering, Brazing, Adhesive Bonding, etc.)	xx	x	xx			x	x	x	
Powder Processing	x	x					x		
Vapor and Electro-Deposition, Epitaxy				x		x	x	x	xxx
Radiation Treatment (Ion Implantation, Electron Beam, UV, etc.)				x			x	x	xx
Plating and Coating	x		x	x		x	xx	x	xxx
Chemical (Doping, Photoprocessing, Etching, etc.)		xx	x	xx		xx	xx	x	x
Testing and Non-Destructive Testing	x				x				

TABLE 15

Priorities for Applied Research in Materials by Area of Impact (con't.)

DISCIPLINES	Communications, Computers, Control	Consumer Goods	Defense & Space	Energy	Environmental Quality	Health Services	Housing & Other Construction	Production Equip.	Transportation Equipment
Earth Sciences									
Analytical Chemistry	x	x	x		x	x			
Physical Chemistry	x	x	x	x	x	x	xx	x	x
Organic and Polymer Chemistry	x	xx	x		x	xxx	x		
Inorganic Chemistry	x	x	x	x	x	x			
Solid State Chemistry	xxx	x	x	xx					
Solid State Physics	xxx	x	x	xx					
Ceramics and Glass	xx	x	xx	x	x	x	xx	x	xx
Polymer Processing	x	xxx	xx			x	x		
Extractive Metallurgy			x		xx				
Metals and Inorganic Materials Processing	x		xx	x	x				
Physical Metallurgy	x		xx	xx	xx	x		x	xx
Chemical Engineering	x	x	x	x	xx	x	x	x	xx
Mechanical Engineering		x	xx	x	x				
Electronic Engineering	xxx		xx	x		x		x	xx
Aerospace Engineering		x	xx	xx		x			
Nuclear Engineering			x		x				
Bioengineering	x	x			x	xxx	x		
Civil and Environmental Engineering	x	x			xxx				

*Due to a typographical error in the original questionnaire, Nuclear and Surface Properties were entered as one item. However, respondents generally read it as Nuclear and included Surface Properties under Chemical and Electrochemical.

TABLE 16

Applied Materials Research Problems of Broad Implication

The combined opinions of a number of materials scientists and engineers suggest upon analysis that high priority be assigned to the generic problems in applied materials research, listed below. These problems are characterized by their broad implications and, for that reason, might well be considered by academic investigators. The problems were selected from among several thousand proposed.

Properties

Chemical: corrosion; stress corrosion; flammability; catalysis.

Biological: biocompatibility; toxicity; allergenicity; biodegradability.

Mechanical: fracture; fatigue; creep; friction; wear; lubrication.

Defects and Microstructure: effects of impurities and crystallographic imperfections on properties.

Electrical: superconductivity.

Materials

Composites: fracture toughness; interfacial phenomena; reliability.

Thin Films: reliability; plating and coating.

High Performance: superalloys; ceramics and glass.

Plastics: property-structure relations; high performance.

Processes

Testing: nondestructive testing; characterization; analysis; interaction with optical, acoustical, and other forms of radiation.

Joining: adhesives; welding.

Polymer Processing: synthesis; extrusion; molding; recycling.

TABLE 17

Priorities for Basic Research in Materials

(x -- indicates above-average priority;
xxx -- indicates highest priority.
See Appendix A for explanation of analysis.)

Rank, Allowing for Familiarity

Chemists	Physicists	Metallurgists	Engineers	PROPERTIES	Uncorrected for Familiarity	Corrected for Familiarity	Experts	Overall Rating
OUT OF 13	OUT OF 13	13						
6	7	7	1	Atomic Structure (Crystallography and Defects)	xxx	xx	x	xx
4	4	3	3	Microstructure (Electron Microscope Level)	xxx	xx	x	xx
13	13	13	12	Microstructure (Optical Microscope Level)	xx	x		x
12	8	9	5	Thermodynamic (Phase Equilibria, Change of State, etc.)	x			
10	12	12	8	Thermal (Thermal Conductivity, Phonons, Diffusion, etc.)	xxx	xx	xxx	xxx
5	9	2	6	Mechanical & Acoustic (Strength, Creep, Fatigue, Damping, etc.)	x	x	xx	xx
9	4	6	9	Optical (Emission, Absorption, Luminescence, Excitation, etc.)				
3	3	8	7	Electrical (Conduction, Electron Trans., Ionic Cond., Thermoelectric, Injection, Carrier Phen.)	xx	xx	xx	xx
8	11	10	13	Magnetic (Ferromagnetic Resonance, Paramagnetic, etc.)				
11	10	11	11	Dielectric (Ferroelectric, Breakdown, Loss, Piezoelectric, etc.)	xxx	xxx	xxx	xxx
7	6	5	10	Nuclear* (Radiation Damage, Absorption, Surface States, Catalysis)	x	xxx	xxx	xxx
2	2	1	2	Chemical & Electrochemical* (Corrosion, Battery Phen., Oxidation, Flammability, etc.)	xxx	xxx	xxx	xxx
1	1	3	4	Biological (Toxicity, Biodegradability, etc.)	x	xxx	xxx	xxx
				MATERIALS				
OUT OF 19	OUT OF 19	19						
3	5	1	5	Ceramics	xxx	xxx	xxx	xxx
6	1	6	4	Glasses and Amorphous	xxx	xxx	xxx	xxx
7	8	7	8	Elemental and Compound Semiconductors	xx	xx	xx	xx
12	11	13	11	Inorganic, Non-Metallic Elements and Compounds	xx	x	x	x
10	16	18	16	Ferrous Metals and Alloys	xx	x	x	x
5	10	14	13	Non-Ferrous Structural Metals and Alloys	xx	x	x	x
13	12	19	15	Non-Ferrous Conducting Metals and Alloys				
4	7	3	3	Plastics				
11	14	11	14	Fibers and Textiles				
14	15	12	12	Rubbers				
1	2	2	1	Composites	xxx	xxx	xxx	xxx
16	6	10	9	Organic and Organo-Metallic Compounds		x	x	x
9	4	8	6	Thin Films	x	xx	xx	xx

TABLE 17

Priorities for Basic Research in Materials (con't.)

	Rank, Allowing for Familiarity				Uncorrected for Familiarity	Corrected for Familiarity	Experts	Overall Rating
	Chemists	Physicists	Metallurgists	Engineers				
Adhesives, Coatings, Finishes, Seals	8	9	4	2	xx	xx	x	xx
Lubricants, Oils, Solvents, Cleansers	15	13	9	10		x	xxx	xxx
Prosthetic and Medical Materials	2	3	5	7	x	xxx		
Plain and Reinforced Concrete	17	17	15	17				
Asphaltic and Bituminous Materials	19	18	17	19				
Wood and Paper	18	19	16	18				
PROCESSES	OUT OF	OUT OF	14					
Extraction, Purification, Refining	2	4	5	8	x	xx	xxx	xx
Synthesis and Polymerization	4	1	3	2	xx	xxx	xx	xx
Solidification and Crystal Growth	8	5	9	3	xxx	x	xx	xx
Metal Deformation and Processing	6	11	12	12	x			
Plastics Extrusion and Molding	13	12	7	10				
Heat Treatment	11	14	14	14				
Material Removal (Machining, Electrochemical, Grinding, etc.)	10	13	13	13	xx	xx	xxx	xx
Joining (Welding, Soldering, Brazing, Adhesive Bonding, etc.)	5	9	2	5	x	x	xx	x
Powder Processing	3	10	4	7	x	x	xxx	x
Vapor and Electro-Deposition, Epitaxy	9	3	10	4		x		x
Radiation Treatment (Ion Implantation, Electron Beam, UV, etc.)	7	2	8	9				
Plating and Coating	12	8	6	11	x		xx	
Chemical (Doping, Photo-Processing, Etching, etc.)	14	6	11	6				
Testing and Non-Destructive Testing	1	7	1	1	xxx	xxx	xxx	xxx

*Due to a typographical error in the original questionnaire, Nuclear and Surface Properties were entered as one item. However, respondees generally read it as Nuclear and included Surface Properties under Chemical and Electrochemical.

given highest priority (three stars in Overall Rating) for basic
research are:

- Properties: biological; chemical, particularly
  surfaces; mechanical.
- Materials: ceramics; composites; glass and amorphous;
  plastics; prosthetic.
- Processes: testing and nondestructive testing.

As with applied research, lists of basic research topics in the various
specialties appear in Appendix A.

Information from the priorities questionnaire also allowed
comparisons to be made among respondents grouped according to their
fields of highest degree. The left side of Table 17 shows the rankings
arrived at in this way by four groups -- chemists, physicists, metallurgists (including ceramists), and engineers -- taking into account
average familiarity within each group for each specialty. The rankings
display both good correspondence and intriguing differences. It
appears, for example, that those who would be expected to know most
about a given specialty sometimes rate it lower than do materials
professionals in the other disciplinary groups. Thus metallurgists
rate the priority of basic research on ferrous metals lower than do
any of the other disciplinary groups; physicists, who have much to
contribute to nondestructive-testing methods and instrumentation, rate
it seventh among processes, while the other three groups rate it first.

A possible interpretation of the rank orderings on the left side
of Table 17 is that they are arranged roughly in accordance with the
degree of opportunity as perceived by the four groups of professionals.
That is, the highest-ranked items are those of greatest scope and need
for generating new knowledge.

These rankings, however, clearly must be interpreted with care. In particular, they should not be taken necessarily to indicate relative increments of needed research support; rather they might be taken to suggest the relative sizes of programs within overall materials research budgets. Nor do the rank orderings show the underlinedexistingunderlined importance of various specialties to the related applied research and engineering. Ferrous metals and alloys, for example, are essential to the economy, but evidently the respondents (even the metallurgical group) felt that basic research in this field might be expected to yield diminishing returns today, perhaps because of extensive research in the past. Materials like concrete, asphalt, and wood, in contrast, have not been subjected to comparable basic research, so that the corresponding fundamental understanding may not yet be advanced to the point where research opportunities are recognizable, even by experts in the field. Yet, in view of the enormous role of the latter materials in the nation's economy and way of life, a modest investment in research could ultimately yield a relatively large return compared with that from many other research areas.

Selected Priority Problems in Materials Research Based on Questionnaire Responses

Corrosion. Although much progress has been made in understanding the thermodynamics and kinetics of the corrosion process, the mechanisms of localized corrosion are not well understood, nor are those for imparting resistance or protection against aqueous or gaseous corrosion.

For localized corrosion like pitting and stress corrosion, initiation is distinct from propagation. Initiation may involve the breakdown of a surface film; important factors to be studied are variations in film composition and microstructure down to the atomic level and their interaction with the environment. Corrosion can be initiated also at surface inhomogeneities, but the types have not been characterized clearly.

The propagation of stress-corrosion, hydrogen-embrittlement, and corrosion-fatigue cracks demands further investigation. As the use of high-strength materials increases, these problems become more important. Susceptibility to hydrogen embrittlement, for example, increases with the strength of the steel. The mechanism of stress corrosion probably differs in detail from system to system. Problems pertinent to many systems include the role of mechanical fracture; the effect of stress on the rate of anodic dissolution; continuous versus discontinuous cracking; the relevancy of continuum mechanics as opposed to atomistic analyses of crack propagation; the effect of defect structure and of chemical composition and distribution at the macro and micro levels in the metal; and the role of hydrogen generated at the crack tip.

High-technology industries often must cope with unexplored conditions. Thus, research is required for corrosion in aqueous media at high temperature, high pressure, or both; in the ocean, near the surface and at great depth; in highly corrosive body fluids for prostheses; and in gaseous media for thin metal films, whose properties may differ radically from those of the bulk material.

The corrosion of alloys in gaseous environments can cause surface roughening, most likely because of preferential attack on the less noble constituent, and can result in poor surface finish and poor adhesion of films. The theory of surface instability and the mechanisms of surface roughening require further attention.

Small changes in chemical composition can radically change the corrosion resistance of an alloy because of subtle alterations in the characteristics of surface films. Research must determine the effects of alloy composition and structure on surface films and relate these effects to the problems of internal and external oxidation, the adhesion and the spalling of corrosion films, their resistance to breakdown, and the mechanism of self-healing. More specifically, the following must be examined: the crystallography of surface films and the factors that determine crystal size and transitions between the crystalline and amorphous state; the defect structures; the conductivity of, and diffusivities within, the films and their effects on film-growth kinetics; the mechanical properties of corrosion films; and the thermodynamics and kinetics of the transformation from one corrosion product to another during the high-temperature gaseous corrosion of complex alloys. Such studies could lead to new alloys with better corrosion properties, or to less costly compositions.

Protective coatings fall into two classes: inhibitors, which are of monomolecular dimensions and reduce the anodic and cathodic reaction rate; and thicker films, which provide a physical barrier. The interaction of inorganic inhibitors like chromates with a metal surface requires elucidation. Are they adsorbed? Are electrons transferred (i.e., is the metal oxidized)? Are all surface sites affected equally? The application of metallic coatings can result in the formation of intermetallic phases at the interface between the metals. Their role in adhesion and in corrosion protection is not sufficiently clear. The possibility of using metals like chromium or aluminum, which form corrosion-resistant oxides, as coatings on the refractory metals for service at high temperature (above 1,200°C) should be studied more systematically, along with the resulting chemical and metallurgical problems. For organic films, basic research is needed on the mechanism of adhesion.

Flammability of Polymers. Flammability, an especially fast form of surface chemical reaction, is particularly important in the use of polymers. The controlling variables of burning must be determined more quantitatively. The oxygen index test, for example, rates the ease of burning of individual materials quite well; counterparts must be developed for entire materials systems or products in terms of end-use environments.

The high-temperature, free-radical reactions of polymer combustion encompass oxidation and pyrolysis in both the flame and the degrading polymer. These reactions can be described at present only in qualitative terms. Learning how the important physical and chemical processes may be slowed or altered by adding various fire retardants is a challenge to research very similar to that posed by catalysis.

The success of empirical efforts to devise better fire retardants for flammable materials has apparently peaked. Recent progress has mainly involved optimizing the forms and amounts of antimony, halogen, and/or phosphorus in particular materials or products. Such treatments, however, are now known to increase greatly the formation of smoke and toxic gases. Inherently nonflammable polymers like the polyimides are available, but they are not economical except in limited, small-volume applications.

The greatest improvements in fire protection will come probably from careful design and engineering to give system-wide, rather than individual material, protection. The age-old, reliable sprinkler system is a simple instance. An excellent example of innovation in this area is the recent experiment in which intumescent insulation alone protected the inside of an entire aircraft fuselage for more than 10 minutes in an inferno of burning fuel.

Biomaterials. The development of materials and devices for use in medicine and surgery is an exciting growth area for materials research. At this interface between the animate and inanimate worlds, many questions must be answered at a most basic level.

Among typical research topics in the field is the surface architecture of biomaterials, including surface energy and changes that can result from contact with body fluids. Examples of such changes would be the development of monolayers of lipids or proteins on the material in question.

Further research is in order on the mechanisms of the degradation of polymers by water, lipids, proteins, and enzymes. Needed, too, are studies on the corrosion of metals and the degradation of ceramics, frequently by hydrolysis. Also important is the converse area, the passivation of metals to make them less susceptible to corrosion and

the development of coatings to protect them. Corrosion protection
for ceramics and polymers demands further work as well.

Another target for research in biomaterials is the mechanism
of bonding by adhesives between metals or polymers and, for example,
hydroxy apatite and collagen. The topic is important in both dental
and orthopedic applications. In dentistry particularly there is
pressing need for an adhesive to seal the margins between the tooth
enamel and a restorative material. Such problems in bonding have much
in common with a general challenge in materials science and engineering
-- the striving for better understanding of phenomena at the interface
between two constituents and of the possible degradation at the
interface.

Glass ceramics based on the calcium phosphate glasses are
potentially useful in areas that include degradable orthopedic devices.
Only a small research effort is under way on these materials, however,
and more seems warranted.

Substantial development has been carried out on pyrolytic carbon
for heart valves, but more work is needed on graphite and carbons,
which are compatible with the human body. Major consideration should
be given to fabrication.

New techniques, such as freeze drying, should be studied for
fabricating materials with potential as implants. Particularly intriguing
also is the replamineform process, which replicates life forms or
structures in the appropriate material. Much remains to be done in
the development of membranes suitable for diffusion of gases. These
are needed for devices that measure oxygen, carbon dioxide, and other

gases, as well as for long-term artificial-lung devices. Measurement of gases is important for in vivo physiological studies on the body. Another significant field is membranes for kidney dialysis. The big problem here is cost, since membranes currently exist that maintain life. As the emphasis shifts toward full rehabilitation of patients, however, there will be a need for a totally implantable artificial kidney of superior diffusion and surface properties as well as acceptable cost.

Reversible physico- or chemi-adsorption requires further investigation. The topic may become central in the area of drug release or its reverse, adsorption of toxic materials. Both polymeric and ceramic or glass systems have potential for such uses. One objective is to be able to "dial in" the release rate, so that the depot material releases the drug into the blood-stream at a preset rate.

Another research area is the application of analytical techniques in monitoring variations in tissue components and fluids. Such an instance would be changes in the conformation of polymers or proteins in the body, such as hyaluronic acid, a polysaccharide which is important in arthritis and is part of the synovial fluid in joints; another would be changes in the proteins collagen and elastin in blood-vessel walls, which is important in atherosclerosis. One would also like to know how conformational and other changes may alter the calcium binding.

The effects of stress on the development of biopotentials in natural tissue is still an open field. Stress is known, for example, to cause changes in the rate of dissolution or deposition of hard tissue.

Blood clotting at implant sites remains an enormous question. A major part of the problem concerns the surface-chemical architecture of the implant material and its effect on protein adsorption. To date, several materials, such as pyrolytic carbon, block copolyether-urethanes, and heparin-treated polymers, have shown encouraging results as anticlotting materials. Each has limitations in use, however. Also demanding attention is the fluid mechanics of blood flow and surface adhesion together with the effect of the implant material in terms of fibrous ingrowth and calcification.

Fracture Mechanisms, Defects. Fundamental understanding of fracture mechanisms is important to the design of safer, more reliable engineering structures. Relatively sophisticated theories of fracture, which take into account polycrystallinity or microstructure, have led lately to better testing procedures and tougher materials. Such improvements in knowledge can conserve materials by minimizing the need for over-design, but much remains to be done in this connection. A relatively new aspect of materials science and engineering in this context is the application of fracture mechanics to rocks and geological structures. Involved here are structure/property relationships on a vast scale, having many implications for tunnel excavating, underground blasting, and seismic damage. Ultimately we have to know more about the actual breaking strengths of atomic or chemical bonds and the role of lattice vibrations in a distorted or defect-riddled structure. Particularly pertinent to the severe technological problem of stress-corrosion cracking is a deeper knowledge of the chemical reactivity of strained interatomic bonds resulting from strained lattice structures.

The first direct observation of dislocations (line defects), in the mid-1950's by transmission electron microscopy, gave great impetus to experimental and theoretical study of dislocations in both single crystals and polycrystalline materials of commercial interest. This work greatly facilitated the analysis of plastic deformation in terms of dislocation motion and interactions. Modern research and development of structural materials is now aided by the concepts and methods of dislocation theory. A current thrust in research on dislocations includes the combination of dislocation theory with continuum mechanics to give a continuum theory of dislocations. Work is needed particularly on the transition from dislocation behavior to continuum behavior under dynamic conditions. Another active area is the use of computers to average individual dislocation reactions into macroscopic plastic deformation.

Dislocation theory is now being applied through analytical models relating the dynamic behavior of dislocations and point defects to mechanical behavior, including such practical applications as creep and hot pressing of crystalline solids. This work is done by computer "mapping," in which various theoretical constitutive equations are used to predict, in stress-temperature space, regions where specific mechanisms of high-temperature mechanical behavior are operative.

The effects of point defects on mechanical behavior are considerable at high temperature, where redistributions of point defects and dislocations may occur. Effects on high-temperature strength are related to vacancy transport (diffusion) and the formation of stable dislocation substructures and networks. Particularly interesting and important is the combined action of high temperature and neutron flux, like that prevailing in a fast breeder nuclear reactor. Here, the dynamics of vacancy agglomeration into voids poses a stiff challenge to research, as does the broader area of gas (hydrogen and helium) generation and agglomeration. The attendant swelling limits severely the practical lifetime of nuclear-fuel elements.

Superconductivity. One of the most tantalizing challenges to materials scientists is to find practical superconductors with higher transition temperatures than those now known. High-temperature superconductors have great potential value in electric power generation and transmission, novel forms of high-speed ground transportation, magnetic ore separation, and many other applications. Hundreds of elemental and compound superconductors have been discovered or synthesized, but the highest transition temperature achieved yet is only 23.2 K (-250°C) in a compound of niobium-germanium. Means of fabricating the compound into useful shapes without depressing the transition temperature have yet to be devised. Indeed, difficulty in large-scale processing is a major obstacle to widespread use of other superconductors having relatively high transition temperatures, such as niobium-tin. Stoichiometry and atomic order are critical material parameters, so that basic research in phase equilibria and kinetics of phase transformations is a necessary prelude to new or improved processing techniques.

On the theoretical side, development of the Bardeen-Cooper-Schrieffer theory in 1957 did much to explain the mechanism of superconductivity and to rationalize various experimental observations. Questions remain, however, on the fundamental limitations imposed on the transition temperature by the lattice and electronic structures of real solids. Much of the search for higher-temperature superconductors has been devoted to finding the appropriate electronic structure, but it has become more apparent lately that the dynamic properties of the lattice (phonons) are at least as important. In particular, lattices that undergo structural transformations accompanied by or triggered by soft modes of the lattice vibrations exhibit some tendency to be high-temperature superconductors. To pursue these clues rigorously and quantitatively is an urgent challenge to materials science. Other recent research results, meanwhile, have stirred excitement in the possibilities of discovering new superconductors that might be lurking in organic materials.

Composites, Concrete. A composite material generally combines two or more mutually discrete macroconstituents -- glass, plastic, ceramic, or metal -- differing in composition or form. Fiber-resin composites usually consist of glass fibers in an epoxy or polyester resin. Many metal-matrix composites have been studied, including steel or boron fibers in aluminum, aluminum oxide fibers in iron, and tungsten fibers in copper or stainless steel. Perhaps the most prevalent composite is concrete, the single most-used man-made construction material.

Fiber-resin composites can be tailored to many needs and are used widely in such applications as filament-wound tanks, automobile bodies, boat hulls, and translucent glazing. Glass fiber is the most frequently used strengthening agent in these materials because of its low cost. But where performance requirements outweigh cost, fibers like graphite and boron, as well as special glass fibers, are finding increased usage, as in space vehicles and aircraft. Fiber-resin composites, because of their temperature limitations, cannot replace metals entirely. It is clear, however, that they can do so to a significant extent, and this may be important as metal resources dwindle.

Research and development is required on fiber-resin composites in four primary areas. One need is improved production technology for fibers and resins to reduce overall cost. This is particularly true for advanced fibers and the higher-temperature resins. The second area is the development of more advanced production-processing methods. Many of the fabrication methods used today for fiber-resin composites are relatively expensive and time-consuming. Further work is required also on new fibers and resins to improve overall properties at reasonable cost. The fourth primary need is advanced design methods and analytical techniques to help the designer exploit the great flexibility of fiber-resin composites more fully. Structural analysis of these complex materials is still in its infancy. Other advances to be sought include better joining methods, higher resistance to erosion by rain and dust, greater toughness, and the development of structural-property data banks. In some uses, problems with flammability may call for work on resins with flame-retardant properties and nontoxic combustion products.

Metal-matrix composites are still in the development stage. Their primary advantages over conventional materials are their high strength and stiffness relative to weight. These materials, in addition, have a good service-temperature range. A number of manufacturers have made and tested metal-matrix composites, often as prototypes.

Prospective applications include aerospace engines, fan and compressor blades in advanced gas turbines, thermal protection surfaces for space vehicles, and a variety of commercial functions.

Concentrated effort is needed to improve the fibers available for metal-matrix composites and to lower their cost. Fibers now under development, including glass, have potential for easy, low-cost manufacturing. Glass-coated aluminum oxide fiber, for example, could have four times the specific modulus of conventional glass fibers and be made by a relatively inexpensive process. The promising high-performance fibers should be combined with metal-matrix materials and studied for mechanical, thermal, and chemical (or metallurgical) compatibility. Equipment should be designed especially for producing fiber-metal matrix composites. And parts or prototypes of components will have to be fabricated and tested under service or simulated conditions to build confidence in these composite materials.

Basic research on concrete is a necessity to help fill the growing demands of construction. The material is a rather complex composite, and the introduction of any single constituent will affect the properties of the others. Thus the problem lies not only in working out the properties of each constituent -- cements, aggregates (sand, stone), reinforcement -- but also in clarifying the interrelationships among them.

Research on physicochemical properties as a function of composition will aid the development of cements designed for specific functions. These include expansive cements for shrinkage control or for self-stressing; cements with controlled setting time; and cements with improved resistance to weathering and physical, chemical, and thermal attack. Challenging problems include studies of the properties of admixtures and their chemical reactions with each other and with reinforcing materials.

The replacement of stone with light-weight aggregates will improve the strength-to-weight ratio of concrete, allowing it to be applied more effectively in tall buildings and long spans. These light-weight aggregates include waste products like fly ash as well as inexpensive natural aggregates indigenous to local construction sites. The physicochemical properties of these materials in relation to the other constituents of concrete require thorough investigation.

Reinforcing materials such as steel rods and fibers increase the tensile strength and toughness of concrete. Studies of adhesion and the reactivity of steel with cements and admixtures of controlled composition are essential to learn to retard the degradation of properties with time. Alternatives to steel, such as organic-coated glass fibers, may broaden the utility of reinforced concrete.

Superalloys. Alloys based on nickel, cobalt, or iron and intended for service above 500°C are frequently termed superalloys. More than 50 superalloy compositions are commercially available in this country, but the nickel-base system is the most widely adopted. Such super-alloys account for some 50 percent of the weight of aircraft gas-turbine engines, where they are used for turbine and compressor disks, turbine vanes and blades, and other hot components. It appears that, to an appreciable extent at least, the ability of engine designers to achieve their plans for advanced engines will depend on progress in superalloy metallurgy.

Alloys developed recently in the United States and Britain have maximum practical use limits varying from 1,000° to 1,050°C. Further improvement in superalloys seems likely to come from two directions: overcoming the temperature limitations based on environmental attack, and increasing the strength by advances in processing methods. There is also an urgent need to improve the correlation between laboratory tests and service conditions so that the service life of components can be predicted more accurately.

The next generation of superalloys will operate at temperatures too high for traditional chromium-oxide protective scales. One potential solution is to develop a family of superalloys protected by aluminum-oxide films. The latter tend to spall during thermal cycles, but this may be inhibited by introducing dispersed oxides into the alloy. It has been found only recently that dispersed oxides are quite beneficial in combating high-temperature corrosion in addition to their favorable effects on high-temperature creep. Major questions to be answered in the development of protective coatings include the influence of alloying additions on the diffusivity of the constituents; on the thermodynamics and kinetics of formation of competing oxide films; on competition between internal and external oxidation; and on vacancy behavior and its possible role in spalling.

Processing offers many possibilities for enhancing the properties and performance of superalloys. Promising approaches include the use of directionally solidified eutectic alloys, electroslag remelting, composite structures joined by diffusion bonding, and improved powder-metallurgy processing. Mechanical alloying in attritor mills can effectively disperse oxide phases. Control of recrystallized structures to yield interlocked elongated grains aligned in the loading direction can be achieved by adjustment of dispersed phases or through zone-controlled recrystallization. The high-temperature benefits resulting from elongated, interlocked grains in nonsag tungsten can likely be extended more broadly to superalloys and other high-temperature materials.

Ceramics, Glass. A major aim of ceramics technology is to improve the physical and mechanical properties of polycrystalline ceramics, which compete with metals and glasses as engineering materials. The general approach is to develop new compositions and processing techniques that permit superior properties to be achieved through close control of composition, density, and grain structure. Thus, increasing emphasis will be placed on the relationships among composition, microstructure, and material properties.

The traditional uses of ceramics rely heavily on the materials' durability and resistance to thermal and chemical attack. There is much opportunity for improvement here, in new facing materials and glazing systems for buildings, in prefabrication materials such as blocks and foams, in brick and pipe with improved weatherability and resistance to frost. The upgrading of the thermal-shock characteristics of ceramics for furnaces and thermal reactors depends on deeper insight into the chemistry of oxide formation and the variables of composition and particle size and shape. Such ceramics are used increasingly in applications such as incinerators and kilns.

For new compositions, basic study in solid-state physics and chemistry should clarify electrical, optical, magnetic and mechanical phenomena that may be peculiar to ceramics. In the recently discovered lead-lanthanum zirconate titanate ceramic, for example, only a small addition of lanthanum improves optical transparency and electro-optic memory characteristics. This suggests that effects of impurities on physical properties of ceramics should be examined more systematically.

High-temperature structural ceramics are receiving new emphasis. One example is the current work on silicon carbide and silicon nitride for service in gas turbines, where they would permit higher operating temperature and thus greater efficiency. Ceramics are not tough enough for this application at present. One approach is to learn to design around the shortcoming through deeper understanding of fracture mechanics. A second direction is to find means of improving the toughness of materials like silicon carbide, which should be a superior high-temperature material, but which cannot yet be fabricated economically with sufficiently close control of its structure.

Recent studies have indicated that improved high-temperature properties may be possible in complex systems such as solid solutions of nitrides and oxides of silicon or aluminum; here, high density can be achieved by sintering at relatively low temperatures. Accordingly, ceramics based on silicon-aluminum-oxygen-nitrogen and related systems are promising candidates for structural materials.

Composite materials containing ceramic fibers represent a way to combine a high-strength fiber with a ductile matrix, but the interface-bonding problem inhibits full realization of their potential.

The limited ductility of ceramics also restricts their use as light-weight armor. The wear resistance of ceramics makes them good cutting tools but, again, their limited fracture-toughness restricts their applicability.

Ceramics with good thermal-shock resistance have been developed for stove tops and have many other potential uses. Lucalox, the transparent lamp-envelope ceramic, was developed through an understanding and control of sintering behavior; it points the way to other transparent ceramics. Important here is knowledge of basic processes: mixing of ceramics; preparing and processing highly reactive starting powders (such as aluminum oxide); control of sintering behavior; and stabilization of properties at high temperatures.

Glass formation is basically a kinetic phenomenon, and much remains to be learned of the associated dynamic problems, including diffusion, conductivity, and polarization. The separation of homogeneous glass into two amorphous phases, or into amorphous and crystalline phases, may be either troublesome or useful, depending on the application. Neither the thermodynamics nor the kinetics of these processes are sufficiently well in hand to allow the occurrence or absence of phase separations to be predicted in the more complex glasses.

A common limitation in the utilization of bulk glass is its brittleness. Investigations of brittle fracture, ultimate strength, notch sensitivity, and static fatigue have led to more efficient use of the intrinsic strength of glass products. The fundamental limits on the mechanical properties of the material, however, remain unknown. The structure of glass is difficult to determine and even to define.

The application of a combination of modern tools like nuclear magnetic resonance, Raman spectroscopy, and electron microscopy should produce further insights in this area.

A new opportunity in glass technology is the development of optical wave-guides for long-distance communications. A commonly envisioned configuration is a high-refractive-index optical fiber with a low-refractive-index cladding. It has been realized recently that, in some inorganic glasses, loss of light in the red and near-infrared spectral regions should be very small. This discovery has produced intense activity in preparing glass fibers of precise dimensions that are extremely pure and free of light-scattering and light-absorbing defects. The work has underscored the need for improved ultrapurification processes for glasses and chemical compounds.

Polymers. A variety of useful plastics, elastomers, and other polymers have become commercial products in recent decades. Launching a radically new polymer, however, is expensive. Semiempirical routes to practical materials are likely to be followed most often, and one of the most attractive is to explore the effects of blending polymeric materials already available. The properties of blends depend on many factors, the most important, perhaps, being the intimacy of the mixing. The degree of dispersion can vary widely. Indeed, some polymer pairs cannot be blended properly at all. Where this is so, various modifications may improve compatibility. One example is acrylonitrile-butadiene-styrene, in which the rubber particles added to increase impact strength are surface-modified by grafting to provide a good bond across the boundary with the base polymer.

A more fundamental understanding of the characteristics of blends is needed to obtain materials with tailor-made properties, properties that are too specialized sometimes to warrant bulk production of a new polymer. Such research on polymer blends can be compared in many ways to the history of research and development with metal alloys.

Major achievement and continuing effort mark the field of rubberlike polymers. A recent product of basic studies is the family of ethylene-propylene copolymers. All rubberlike substances lose their viscoelastic properties and become far more rigid when the temperature is low enough. They assume a glassy character and behave as almost perfect elastic solids. A host of polymers are glassy at ordinary temperatures and many of them, like the inorganic glasses, have valuable optical properties. Fundamental work with monomers or combinations of monomers to yield polymers with desired properties, such as a particular optical absorption or refractive index, has been highly productive and can be expected to be so in the future.

Ability to control the structural regularity of polymer chains has been a striking achievement in polymer science. This chemistry of molecular shape (stereochemistry) is making it possible to synthesize highly ordered molecules that cluster into crystalline order. The individual crystalline regions are extremely small but highly organized. They form a superstructure or morphology that gives strength and dimensional stability to the polymer, somewhat like the way in which precipitates can strengthen metal alloys. The morphology is extremely complex, but is governed nevertheless by identifiable factors, such as the rate of the crystallization and the distribution of molecular sizes. Further research in this area will lead undoubtedly to new and serviceable materials.

The relations between molecular structure and physical properties are central to the behavior of polymers of biological interest. The proteins responsible for form and strength in much of living matter, notably collagen and keratin, are examples of substances in which these relationships are becoming well understood at the molecular level. Continued research in this area is likely to yield both nonbiological and biological uses for plastics.

The durability of polymers may become one of the most active areas for polymer research in the immediate future. Various stabilizers are added to protect commercial polymers against ultra-violet light, heat, and other kinds of degradation. Current studies on polymer durability center on stabilizer interactions, retention, and lifetimes under various conditions. Minor structural modifications have been found recently to improve stability markedly in polyvinyl chloride and polyoxymethylene without significant changes in physical properties. Further increases in stability probably can be expected from additional changes in the molecular structure of polymers.

Processing: Metals. Two main goals of innovation in metal processing are to improve mechanical and physical properties and to make finished parts more economically. Most new processing techniques are developed for specific materials and applications, but basic research in this connection should spawn new approaches.

Melt spinning, a relatively new means of casting metal filaments by extruding a liquid jet through a fine orifice, may lead to high-speed production of fine wires in a single step. Pertinent research problems include the hydrodynamics of the liquid jet and the chemistry of surface films developed to stabilize the jet. Another new technique, rheocasting, involves casting metal that is partially solidified; high fluidity is maintained by vigorous mechanical stirring. The lower pouring temperature reduces mold erosion, centerline shrinkage, and freezing time. If the stirring is stopped momentarily, the slurry stiffens and can be handled like a solid for die casting (thixocasting). Research problems here include fluid flow and rheology of partly solidified alloys and the microstructure and properties obtained in this type of casting.

The properties of practically all commercial alloys, as well as those of new alloys, can be improved by controlling the thermal and mechanical cycles of processing. Progress in thermo-mechanical processing will come from restudy of the complex interaction of deformation, recrystallization, texture development, and solid-state reactions in the important commercial alloys. In addition, the principles of property improvement by thermomechanical processing are rather well understood in many cases, but economical forming methods or systems have not been developed to reduce the principles to practice. Instead, the processes have been adapted to existing facilities, usually with little success. The stiffness of steel, for example, could be increased perhaps one third if the appropriate texture could be produced in polycrystalline iron under production conditions.

One of the greastest opportunities for materials development lies in the use of powder metallurgy and powder consolidation to shape materials and components that cannot be formed by conventional techniques. The possibilities here need to be clarified and reduced to practice. The results should lead to more economical utilization of materials.

In joining methods like diffusion bonding, which is used to fabricate metal-matrix composites, progress requires further knowledge of adhesion as influenced by solid-state reactions under conditions that include heat, pressure, and surface films. For the newer welding techniques -- plasma arcs, electron and laser beams -- structural changes and the resulting properties in the region of the weld require study.

Work is needed also on the changes in structure and composition produced by finishing operations such as electric discharge machining, electrochemical machining, and laser machining. New approaches to coating -- flame spraying, for instance -- would benefit from research on the resulting microstructures and properties. Particularly important is the development of alloys with a built-in ability to generate protective coatings during service. One example is the incorporation of aluminum, chromium, and yttrium in nickel alloys to provide oxidation resistance without an external coating.

"Splat cooling," in which molten alloys are shot onto a cold surface, has created vast possibilities for new materials. The technique quenches the melt so fast that it solidifies with a minimum of atomic diffusion. This leads to the formation of metastable (marginally stable) phases, characterized by a range of unusual properties. The equilibrium form of Nb_3Ge, a compound of niobium and germanium, for example, becomes superconducting at 7 K; splat cooling yields a metastable form of the compound whose transition temperature is 17 K. Splat cooling also produces metallic glass alloys that, mechanically, are among the strongest of the nonferrous materials.

Research on metastable states in the past has concentrated on structure, as in demonstrating the amorphous nature of the metallic glasses. Emphasis now is changing to the use of splat cooling to enhance specific properties of materials or to create new properties. Among many targets for study are the mechanical, corrosion, and transport properties of materials like metallic glasses, which have no grain boundaries or similar imperfections. The unusual properties of splat-cooled materials cannot be fully exploited, however, unless the difficult problem of fabricating them into practical forms can be solved.

Processing: Rubber, Plastics. Many methods have been examined to reduce the relatively high cost of processing conventional rubbers. Some rubbers are now sold as powders that can be mixed initially by blending and then fed directly to an extruder or injection-molding press. An attractive possibility is to mix a low-molecular-weight rubber as a liquid, thus avoiding the power-consuming shearing action required with solid rubbers. After mixing, the rubber could be chain-extended and crosslinked to give a product equivalent to that made by present methods. Advances required to achieve this end embrace a range of elastomers of proper reactivity for the chain extension and crosslinking, as well as the corresponding linking agents.

Recent developments in block copolymers have created elastomers in which the crosslinks are physical in nature rather than chemical and may be formed and broken reversibly by heat. These new rubbers, potentailly, could be processed as plastics. The high creep rates of this class of materials currently exclude them from many applications, but commercial rubbers could result from proper choice of monomers and the development of the necessary polymerization techniques.

Cold forming of both amorphous and crystalline polymers is an active field at present. The method avoids the energy consumption and time required to heat a polymer above its softening or melting point and then recool it to room temperature. Injection molding of thermosetting resins also looks promising. Cycle times for heavy-section moldings are now faster than for injection molding.

A further need in polymer processing is for more efficient recycling. (Biodegradation techniques for plastics are poorly developed thus far.) Reclamation of rubbers by mastication and of the monomer from polymethyl methacrylate by pyrolysis are well established processes, and a method has been developed for recovering polyethylene terephthalate (polyester) from textile mill tailings and photographic film. Means must be found of using polymer scrap as a raw material for new processes. Thus, chemicals can be obtained from automobile tires by destructive distillation, and carbon black by controlled combustion. Likewise inviting attention are recycling processes for polymers that emit unpleasant or poisonous fumes when burned. Polyvinyl chloride is perhaps the most important of these because so much of it is made. When burned, the plastic yields hydrochloric acid, which must be removed from the combustion gas before it is discharged. The problem is to find a process that avoids the need for burning.

Most polymer-recycling processes require a substantial supply of clean scrap, and factory scrap is the first choice. Really efficient operation will depend on adequate separation of plastics from rubbish and garbage along with economical transportation to recycling plants.

Testing, Characterization, Evaluation. The practice of testing and delineating the characteristics of materials, especially those related to performance, runs through all technology. Testing is required for quality control; for establishing standards to ensure in-service durability, reliability, and safety; for sensing in production processes and automation; and to avoid environmental degradation.

In a 1967 report, Characterization of Materials,* the Materials Advisory Board stated, "Attempts to provide the superior materials that are critically needed in defense and industry are usually empirical and often wasteful of efforts and funds. That is so, chiefly because we do not yet have a fully developed science of materials that affords predictable and reliable results in devising and engineering new materials for specific tasks." A definition was proposed -- "Characterization describes those features of the composition and structure (including defects) of a material that are significant for a particular preparation, study of properties, or use, and suffice for the reproduction of the material."

---

* Publication MAB-229-M, National Academy of Sciences - National Academy of Engineering, Washington, D. C., 1967.

Destructive or nondestructive tests are required to determine the many characteristics of materials: mechanical, electrical, optical, and other physical properties; composition and structure; defects and impurities. Understanding of the relationships between properties and performance, of the mechanisms of degradation and failure, and of the interaction of matter with various forms of radiation is essential to the development of testing methods and equipment. The latter, in addition, must be designed to function in the pertinent service environment.

Technology and basic research interact strongly in the development of instrumentation. The initial models of many sophisticated instruments are built, as a rule, for specific research projects. Often this instrumentation eventually becomes standard for production or quality control. One example is the thermocouple, which resulted from basic research in the 19th century on the thermoelectric effect. The thermoelectric properties of many materials were determined, and this led to the adaptation of the phenomenon to measure temperature. Other well-known examples are x-ray diffraction, the optical and electron microscopes, and spectrochemical analysis.

The realization that the composition of the surface of a solid usually cannot be inferred from measurements of the bulk material has stimulated the development of new spectrometric instruments for surface analysis. Much of the current effort is aimed at establishing the full potential of these tools, which include the ion probe, the x-ray photoelectron spectrometer, the Auger spectrometer, and the ion-scattering spectrometer.

Analysis of ultrapure materials is challenging analysts seriously. Improvements are required in the mass spectroscopy of solids and in activation analysis. Ecological concerns are largely responsible for an upsurge of interest in the detection of organic compounds, such as those present in trace amounts in biological materials. Advances will be sought, as a result, in mass spectroscopy, infrared techniques, gas chromatography, electrophoresis, and other analytical methods.

Nondestructive testing is among the areas of materials technology requiring urgent attention. In the past, nondestructive testing generally meant testing only for geometric size, defects, and some mechanical properties, but it should be interpreted much more broadly -- testing for composition, microstructure, and the full range of physical properties. Basic research in solid-state physics and chemistry, aimed at detecting and understanding certain properties of materials, has spawned many of the modern techniques for nondestructive testing. The methods depend heavily on the interaction of matter with optical, electromagnetic, acoustical, and other forms of radiation. A few examples of valuable current techniques, or techniques being developed for nondestructive testing are: electron paramagnetic resonance (fracture of polymeric solids, stress analysis); nuclear magnetic resonance (chemical analysis); Mössbauer spectroscopy (surface-chemical and phase analysis, stress analysis); optical correlation (surface distortion); infrared spectroscopy (thermal analysis, flaw detection); microwave attenuation (moisture content); optical and acoustical holography (stress analysis, flaw detection); acoustic emission (flaw detection).

Routine use of these methods in nondestructive testing, however, requires more understanding of the physics of the phenomenon involved, its quantitative relationship to the physical property to be monitored, and the limits of applicability. Required also is instrumentation that offers improved signal detection and reliability as well as greater physical ruggedness and ease of testing, especially in portability and automatic readout of easily interpretable data.

Materials Research on Fundamental Properties

Historically, most new materials or properties have been worked out or discovered by empirical methods. Rarely indeed is a new material or property predicted from basic principles. An outstanding exception lies in single-crystal materials, especially those used for solid-state electronics. Scientists have achieved a degree of understanding, of the simpler crystals at least, that often allows them to prescribe in advance the compositions that will have the generally desired properties.

This progress has come largely because the single-crystal state of matter lends itself to theoretical analysis, particularly of electronic properties. Only recently have basic scientists begun to turn to the more complex forms of matter, the glassy, polycrystalline, and polymeric states found in most practical materials, particularly those employed structurally.

The urgency of fundamental knowledge will vary in different parts of materials science, and priorities will have to be set. Nevertheless, it would be unwise to conclude that even the most esoteric work will not prove useful in the future. The engineer often wishes to produce practical results in a relatively short time, say one to five years. But usually he will not know beforehand what areas of materials science he will have to draw on. If the knowledge is to be ready when the engineer wants it, scientists may have to be working five to 20 years ahead.

To illustrate, in the early 1950's, efforts to calculate the electron-band structures or energy distributions in crystalline semi-conductors would have seemed remote from the everyday task of trying to make practical diodes and transistors. But such calculations, improved by the relatively large computers that were appearing then, have led to strikingly detailed insights into the electronic and optical properties of semiconductors. The calculations were steadily refined, particularly for semiconductors, and extended to other crystalline materials. Today, phenomena like the bulk negative-resistance effect in gallium arsenide (the Gunn effect), laser action in gallium arsenide, infrared photodetection in semiconductors, and light emission from junctions in gallium phosphide are understandable in terms of the detailed band structures of these materials. And the esoteric results of research of nearly two decades ago in theoretical solid-state physics are nowadays the starting point from which the electronic engineer can embark on specific short-term development projects. The status of similarly fundamental research, described selectively below, will be of special interest to scientists and engineers working in the field of materials.

Selected Research Frontiers

Interatomic Forces, Chemical Bonding, Lattice Stability. There is
no more basic property of a material than its very existence. Yet,
despite enormous advances in solid-state theory, we are unable to
predict from first principles and the appropriate atomic wave functions
the configuration and dimensions of any crystal lattice except for a
few very simple materials. Band-structure calculations have reached
a point where the electronic properties of many crystals can be calculated with remarkable precision, given the crystal structure and
atom spacings. But the fundamental challenge remains -- to relate
the properties of individual atoms to those of a crystalline solid
composed of such atoms, particularly the imperfect solid. Ideally
the goal of research in this area should be to predict the conditions
under which the material forms, its structure, its stability, and its
electronic, chemical, and mechanical properties. But stating the
problem this way makes us realize just how primitive is our
quantitative knowledge of such basic matters as interatomic forces,
chemical bonding, and configurational interactions.

Besides the need for theoretical progress, we shall continue
to depend on sensitive experimental determinations of such basic
descriptions of the solid as the band structure, the phonon spectra,
and the Fermi surface with which to test the soundness of theoretical
calculations. Other experiments are required to provide parameter
inputs for these calculations, such as measurements of intermolecular
potentials and charge distributions, and computer simulations of
molecular dynamics. Meanwhile, to fill the immediate needs of the

materials scientist, efforts should be exerted to provide the best available theoretical descriptions of the _imperfect_ solid (e.g., stacking-fault energy, stress fields around vacancies, impurity atoms, and dislocations). A fruitful approach has been the computer modeling of the defect lattice, using interatomic potentials. Much of the groundwork for cooperative experimental-theoretical progress in these areas appears to have been laid, and advances in the understanding of the basic properties of various materials can be expected to emerge steadily in the coming years.

Microscopic Understanding of Phase Transitions. Although the equilibrium crystal structure can be calculated for a few simple materials, we still lack the fundamental knowledge to predict from first principles the changes in crystal structure that occur with variations in temperature, pressure, or composition. And melting, perhaps the most dramatic phase transition of all, is still largely a mystery from a basic point of view. If we understood melting properly, we would have much more insight into the roles of interatomic forces, cooperative interactions, and related phenomena in determining the structure and stability of solids. The microscopic mechanisms that bring about phase transitions are an object of intense research at present, and part of the deep atraction of this study lies in the remarkable universality of certain general phenomena in the vicinity of higher-order phase transitions, regardless of the type of material or even of the type of phase transition -- the liquid-gas (at the critical point), ferromagnetic, ferroelectric, local order, and superconducting phase transitions are very similar in certain rather profound ways. In each case the thermally driven fluctuations in a particular variable become correlated over increasingly longer ranges as the transition is approached, and the time scale of these fluctuations increases markedly. Some consequences of this are the increase in magnetic susceptibility and dielectric constant at the ferromagnetic and ferroelectric transitions, respectively, and critical opalescence at the liquid-gas critical point. While quantitative correspondences between various transitions have been established (the scaling laws), a true microscopic understanding of phase-transition mechanisms has not been achieved. This is a very important challenge for materials research and for solid-state physics in particular.

In martensitic transformations, for example, there is a large body of evidence that special nucleation sites are necessary for initiating the reaction. The nature of these sites (possibly defect arrays) and the mechanisms of interface propagation during transformation have not been clearly determined. It is equally challenging to develop deeper insights into the mechanisms of phase-transitions in interacting systems, such as the coupling of electron spins and phonons near the magnetic transition.

Amorphous, Disordered State. Less is known about the glassy or amorphous state of matter than about crystalline matter. But the possibility that the completely disordered state offers the next most tractable model after the perfectly ordered or crystalline state is attracting widespread theoretical and experimental attention. A fully developed conceptual framework for amorphous materials is lacking, and close collaboration between experimentalists and theorists, as recommended recently by a panel of the National Materials Advisory Board[*], is a prerequisite for major progress to occur. On the theoretical side there is need for calculation of electron-energy diagrams, the counterparts of the energy-band diagram of the crystalline state, that will further understanding of the electrical and optical properties of physically realizable glass structures. In addition to continuing debate over the detailed changes occurring in bond structures on passing from the crystalline to the glassy state, there are questions concerning possible electronic phase transitions and high electric-field transport effects in semiconducting glasses. The correct interpretation of optical-absorption and photoconductivity spectra in terms of electron-energy states is by no means clear; nor are the mechanisms of the converse radiative recombination transitions that give rise to luminescence.

The experimental approach calls for better understanding of material-preparation variables and the glassy-to-crystalline transition; the effects of illumination on the kinetics of this transition seem particularly intriguing as well as possibly lending themselves to various optical writing and memory applications. In the same vein, more needs to be known about the photochemistry of glasses, that is, changes in the electronic states of impurities or imperfections as a result of irradiation with light. The characterization of the effects of radiation damage in an already disordered system have to be unraveled. What determines the mechanical strength and other physical

---

[*] Fundamentals of Amorphous Semiconductors, National Academy of Sciences, Washington, D. C., 1972.

properties of glasses of various compositions and bond coordinations remains an important unanswered question. Indeed, it appears that we still do not comprehend properly the structural states that glasses may assume -- recent work on ultrasonic attenuation at ultralow temperatures may help answer some of these questions.

Impurity Effects in Solids. When impurities are introduced into an otherwise perfect host crystal, all of the properties of the resulting system, in principle, are modified. The nature and extent of these effects depend on the impurity concentration, location, and interaction with the host material. In dilute amounts some impurities can be viewed as a nonperturbing probe of the microscopic properties of the host (as in spin-resonance experiments), but in high concentrations they can lead to new phases (alloys) and phenomena (e.g., order-disorder transitions). Impurities may be desirable, as in most semiconductor phenomena, or undesirable, as in impurity-enhanced optical damage in nonlinear optical materials. Despite the enormous amount of work that has been done on impurity effects in semiconductors, for example, we lack a general theory of the effects of impurities on material properties at the microscopic level. The dilute limit, while theoretically simplest, is experimentally difficult, while the converse is true for high concentrations of impurities.

These problems present several points of attack. The intermediate domain, in which impurity-impurity interactions are no longer negligible, constitutes a prime challenge to both theory and experiment. Recent experiments have shown the existence of cooperative impurity modes (such as phonons, excitons, and magnons) at intermediate concentrations. Theories are needed to explain the emergence of these phenomena from the single-impurity behavior at low concentrations and to distinguish them from the solid-solution behavior at high concentrations. Experiments are in order on systems in which the impurity-host interactions are sufficiently weaker than impurity-impurity interactions to compensate in a controlled way for the numerical abundance of the former. More consideration should be given to systems with simply structured and/or inert hosts, such as helium and the other rare-gas solids, so as to provide theoretically tractable, experimentally accessible model systems for impurity effects. The possibility of long-range order (e.g., magnetic) in the impurities, but not in the host, is particularly intriguing. Progress along these lines has been made already in thin-film studies of magnetic impurities in nonmagnetic, metallic hosts. Similar experiments on optical and electrical properties appear very promising. A lingering puzzle is the role of impurity excitons (electron-hole pairs) in semiconductor laser action. Optical studies have suggested the presence of excitonic molecules and have stimulated speculation on the possibility of creating, within a crystal, a fluid or perhaps even a solid phase composed entirely of electron-hole pairs.

Another continuing controversy concerns the role of interstitial impurities in increasing the low-temperature yield strength (thereby enhancing brittleness) of body-centered cubic metals. One school argues that the lattice-friction stress of the pure metal is inherently large at low temperature, while another contends that interstitials introduce lattice distortions, which are especially effective in impeding dislocation motion at low temperatures. Although sophisticated experiments are required here, theoretical calculations based on interatomic forces could help decide the issue.

Surfaces. Surfaces and interfaces are possibly one of the most fruitful research topics in materials science. Knowledge at the most fundamental level in this area can be expected to be relevant to almost all uses of materials, from the processing and performance of integrated circuits to the corrosion of structural components, from frictional wear and energy loss to catalysis and flammability, from crystal growth to adhesion. The variety and complexity of surfaces and surface layers are at least comparable to the variety and complexity of bulk properties, but our understanding of surfaces is, in contrast, in its infancy.

The aim is to develop more sophisticated insight into the electronic and chemical properties of surfaces. These properties are very sensitive to the detailed ways in which atoms are positioned at the surface, however, and in general these positions are not known. Surface properties are related also to the properties of the underlying bulk material, but in ways that are not often clear. And though bulk properties, by-and-large, are understood in principle, if not always in detail, this is not true of many of the surface properties, where the broad outlines of the phenomenology are only now being drawn. This phenomenology concerns, for example, the details and statistical mechanics of surface topology, local bond and electronic structures, the energy states of electrons at surfaces, and models for nucleation and growth.

Surfaces offer an extra degree of freedom for the arrangement of atoms statistically on the lattice sites. The statistical mechanics of this situation, extending in three dimensions over several atomic layers, needs considerable development. The roughness of a surface on the atomic scale has a major impact on adsorption, surface diffusion, and crystal growth, but very little is yet known about the detailed role of surface roughness in these processes.

The electronic properties of surfaces in simple systems warrant considerable attention. There is some controversy about the extent to which surfaces can be treated as an extension of the bulk -- that is, whether the discontinuity in properties at the surface is great enough to require new concepts and analytical procedures. Our theoretical models for surface electronic properties, surface relaxation, and surface structure are rudimentary. The extent to which surface states on semiconductors are intrinsic to the surface or associated with surface impurities is under debate. Surface states occur both at free surfaces and at interfaces, such as the silicon-silicon oxide interface. It has been shown recently that various surface states on semiconductors correlate with various surface structures as revealed by low-energy electron diffraction.

Surface nucleation, vapor deposition, adsorption, and surface contamination, topics with clear practical significance are currently being investigated in detail for a variety of systems, with emphasis on the simpler systems. The kinetic and thermodynamic properties of vapor deposits can be obtained by mass spectrometric methods, and the distribution of clusters on the surface can be determined by diffraction methods. Classical surface nucleation theory is inadequate to account for the results of such measurements, and major modifications of the theory appear to be necessary. Adsorbed atoms can be identified by Auger spectroscopy, even at a small fraction of a monolayer coverage. Auger spectroscopy coupled with ion bombardment can be used for profiling, to get at bulk composition profiles below the surface. Low-energy electron diffraction is just entering the stage at which the position of surface atoms can be determined quantitatively with some accuracy. These methods are also being used extensively to monitor the cleanliness and structure of surfaces and to investigate production problems involving contamination at surfaces. The electronic and chemical properties of surfaces and adsorbed species are being investigated by a variety of methods. Photoelectron spectroscopy and ultraviolet photoemission spectroscopy are used to obtain band-structure data. Knowledge of electronic and chemical bonding can be derived from ion neutralization spectroscopy. Infrared reflection spectroscopy gives information about chemical bonding, and insights concerning deep electronic levels can be obtained from the analysis of Auger spectra.

The techniques developed for surface research, such as ion mass analysis and Auger spectroscopy, are providing the best, and often the only, methods for investigating materials problems associated with thin films, grain boundary segregation, interdiffusion phenomena, and trace analysis. The trend toward miniaturization in electronics, resulting from economic, reliability, and high-frequency considerations, points toward growing importance of surfaces. The concepts of miniaturization are best embodied in the technology of large-scale integrated circuits, where surface and grain-boundary diffusion often dominate bulk diffusion processes. This trend is expected to continue, particularly as optical microcircuitry is developed.

The elucidation of catalytic processes is not detailed in most cases. Considerable qualitative insight is available, but the roles of surface structure, surface defects, surface geometry, surface electronic properties, and even the bulk properties of catalysts have not been clarified in detail.

Notable advances have been made in the area of adhesion, where knowledge of the role of adlayers and their interaction has contributed significantly. Friction is understood in some detail, especially the role and interaction of the asperities in sliding contact, but the process is difficult to treat from a fundamental standpoint, let alone circumvent in practice. From a practical point of view, the lubrication of sliding contacts is fairly well understood, but cold welding can be a serious problem in electrical contacts. Erosion, corrosion, and contamination of electrical contacts as a result of arcing remain serious problems.

Deeper knowledge of the behavior of surfaces can also be expected to improve our control over the important practical problem of corrosion -- the interaction of a metal with its environment. The presence of water or an electrolyte solution changes the physics and chemistry of metal surfaces significantly. The surface energy is altered and becomes a strong function of the charge in the electrical double layer at the metal/solution interface. The equilibrium surface structure may be different from that in the presence of the metal's own vapor or in a vacuum, and it presents extra problems because the interface is not readily examined in situ. Some metals, such as silver, undergo surface rearrangement in aqueous solution at room temperature. Alloys generally undergo a change in their equilibrium or steady-state surface composition. The atomistics of these phenomena are poorly defined. There is much ignorance regarding the effects of surface stress, defect structure, and nonequilibrium conditions on the reactivity of metal surfaces, and these effects are of major importance in the performance of materials.

One- and Two-Dimensional Systems. Until recently, calculations of physical phenomena in one- or two-dimensional systems were considered to be mainly of academic interest. Onsager's famous exact solution for a simple two-dimensional lattice inspired solid-state physicists and engendered hope for eventual similar success in three-dimensional systems. Within the past four or five years, however, a variety of magnetic, superconducting, and resistive materials have been prepared that exhibit exceedingly large anisotropies in their thermodynamic, transport, and collective properties. (See subsequent section on collective behavior.) The anisotropies are so pronounced that microscopic interactions along a line or within a plane may be several orders of magnitude greater than in the transverse directions. Tetragonal crystals of the K_2NiF_4 family, for example, exhibit inplane magnetic exchange forces several thousand times larger than the out-of-plane exchanges, with the result that below about -170°C truly two-dimensional long-range magnetic order occurs. Neutron diffraction and optical experiments have confirmed the two-dimensional nature of the electron-spin dynamics (magnons) and the critical behavior as well. Similar striking behavior in one-dimensional antiferromagnetism has been observed. Layered-structure transition-metal dichalcogenides (MoS_2, etc.) have long been recognized as effective lubricants. They have now been found to be essentially two-dimensional superconductors, whose properties can be altered markedly by chemically changing the spacing between layers. Certain organometallic complexes have exhibited one-dimensional manifestations of antiferromagnetism and the metal-insulator transition.

The recent evidence of unusually high electrical conductivity in some crystals made up of organic molecules (abbreviated to TTF/TCNQ) has excited considerable interest in the possibility of high-temperature superconductivity in such materials. Whether the high conductivity in fact is related to superconducting phenomena has yet to be demonstrated. But whatever the origin of the effect, if it is real it is a major breakthrough in the properties of organic materials.

These discoveries have kindled lively theoretical and experimental interest in the physics of less than three dimensions. The consequences of extreme anisotropy of microscopic interactions must be explored more fully. The effects of lower dimensionality on collective modes, e.g., electron and heat transport, must be clarified. Particularly intriguing is the effect of a microscopic upper limit to the interaction distance in certain directions on the critical properties near phase transitions in lower dimensional systems. While some magnetic transitions have been studied in this context, virtually nothing has been done on structural, order-disorder, or ferroelectric transitions in less than three dimensions. Further advance in the physics and chemistry of two-dimensional systems is also essential to the eventual understanding of catalysis. Because of the extreme anisotropy in bonding strength in the layered-structure materials, study of their mechanical behavior could lead to superior lubricants or high-strength components, as demonstrated already in graphite. In the usual powder form, graphite is a widely used lubricant. Precursor polymer filaments can be processed to yield dense, highly oriented graphite fibers that exhibit axial strengths that are a significant fraction of the theoretical strength.

Because in some ways they are fundamentally different from bulk materials, thin films and filaments are of renewed interest to solid-state physicists. The fabrication of structures that extend only a few tens of angstroms in one or two directions has made clear the opportunity for more careful experiments and sophisticated inter-pretations in the physics of such structures. Two indicative examples are the observation of a nearly fivefold increase in the superconducting transition temperature in thin films of aluminum and the increased sound-attenuation coefficient of small-diameter glass fibers.

Physical Properties of Polymeric Materials. Polymeric substances, whether natural (such as cotton, wool, and silk) or synthetic (including rubber, rayon, and celluloid), owe their remarkable physical and chemical properties to the long-chain molecules of which they are composed and which set them apart from a host of other materials. Recognition of the key role of long-chain molecules was one of the singular discoveries of this century. It led to intense research to find how variations in the structure of these giant molecules, through new approaches in chemical synthesis, could be invoked to cause valuable changes in the physical properties of plastics. Yet, to put the structure/property relationship of polymeric materials on a firm, fundamental, and quantitative basis remains a prime challenge to materials research, even greater in complexity than the parallel challenge posed by amorphous inorganic materials.

In polymeric materials the molecules may be arranged in an orderly chain-folded fashion; in this form plastics bear some correspondence to the familiar inorganic crystalline materials. But more often the molecules are arranged in a haphazard fashion, resembling perhaps a bowl of spaghetti; this is the counterpart of the disordered, glassy state of inorganic matter. And as with inorganic glasses there can be partial devitrification in plastics. In view of the primitive state of theoretical concepts and analytical procedures for dealing with ordinary glasses, it is not surprising that we are a very long way from being able to go the whole distance of determining from first principles the fundamental properties of the polymeric molecules themselves and then the physical properties of the macroscopic plastic materials.

Collective Behavior. One of the most useful concepts in solid-state physics is that of the collective mode, that is, a simple excitation of a system of interacting electrons and/or atoms. This concept has permitted the handling of complicated many-body (10^{23}) systems in terms of a very few degrees of freedom. The basic idea is to regard the structure and composition of the system as given and to seek its responses to various types of disturbance. The complete set of these responses forms the so-called "normal modes" or "elementary excitations" of the system, which provide the basis for many of its static and dynamic properties. Since a single elementary excitation involves the participation of all the atoms in the system, the concept is quite powerful in elucidating the cooperative behavior among large numbers of particles that results in a particular phenomenon or property. As was indicated briefly in the discussion of phase transitions, the collective-mode concept has been fruitful in describing even anomalous material properties. Although the elementary excitation concept has become very familiar to physicists (the words phonon, plasmon, magnon, etc., are well incorporated into the solid-state vocabulary), it still has great potential for significant growth. Extensions of the concept should prove valuable in at least two directions: (a) nonlinearities and interactions among elementary excitations; and (b) elementary excitations in systems lacking long-range order.

(a) Recent experimental advances have permitted fairly direct and precise study of the more familiar excitations on the one hand, and the generation, detection, and study of some new excitations on the other. In the former category are inelastic scattering (both light and neutron) and acoustic, magneto-optic, and certain solid-state plasma experiments. The latter include super high-frequency phonon and second-sound generation by electron-pair deexcitation in superconductors; the launching of stable-amplitude pulses of both mechanical (e.g., solitons) and electromagnetic (e.g., self-induced transparency) nature; and propagating electroacoustic domains in semiconductors. For the future, better understanding can be expected of the interactions among these excitations, leading to optimized manipulation of such interactions for energy or information transfer.

(b) Less straightforward, perhaps, but certainly no less important is the second direction: studies of elementary excitations in systems lacking long-range order. In amorphous solids and liquids, effort of this kind has been under way for some time. Already, for example, some microscopic understanding of electronic, optical, and acoustical properties of such materials has emerged. Recent generalizations of the hydrodynamic equations to shorter-length and higher-frequency domains have revealed the smooth transition from collective, phonon-like behavior to diffusive and even single-particle behavior in liquids. Some of these trends should also be evident in visco-elastic solids, but the picture is not yet clear. Similar mathematical techniques have been employed to describe elementary excitations in the paramagnetic (disordered) phase of a spin system. The collective modes of the liquid-crystal state are under investigation and should illuminate that important intermediate regime between well-developed long-range order (crystal) and the more transient short-range order (liquid).

Another attractive possibility lies in extending the collective-mode concept to large but finite structures, particularly to macro-molecules. From the point of view that a large molecule approximates a small solid, the existence of collective motions within the molecule is clear. However, the detailed nature of such excitations and their role in transport of charge, strain, spin, etc. within the molecule remain as unusual challenges to both theorist and experimentalist in solid-state physics. A true science base for "molecular engineering" rests largely on progress in this direction.

Nonequilibrium Systems. Basic understanding of the physics of materials under equilibrium conditions is far ahead of that for nonequilibrium systems. While the reasons are not hard to find (such as the limitations of thermodynamics and statistical mechanics), the increasing importance of nonequilibrium phenomena requires that substantial effort be directed to alleviating these deficiencies. Lasers and negative-resistance semiconductor devices are familiar examples of nonequilibrium physics in action. Recent progress in clarifying the transient and threshold behavior has illuminated analogies with equilibrium higher-order phase transitions. It is intriguing to consider more general instabilities such as hydrodynamic, magnetohydrodynamic, and plasma phenomena from this point of view. The problem of turbulence may be the most challenging and important of these. Autocatalytic chemical reaction systems give rise to large spatial and temporal variations in composition. The familiar convective instability can cause extreme problems in crystal growth from the melt. Indeed, the behavior of the atmosphere, the oceans, and even of the earth's crust is strongly influenced by such hydrodynamic instabilities.

With new laser techniques, materials under extreme transient conditions (shock waves and high electric, magnetic, or optical fields) can be studied in real time with a resolution of $\sim 10^{-12}$ second. Scattering, absorption, and fluorescence experiments, which have proved so valuable in guiding theories of materials at equilibrium, should begin soon to do the same for nonequilibrium systems. A foretaste of what might be in store is the use of these fast laser pulses to study short-lived excited states of radicals and molecules, with consequent insights into the detailed sequence of atomic or molecular events taking place in chemical reactions.

RECOMMENDATIONS - AN AGENDA FOR ACTION

In the course of its study, COSMAT has found that the concept
of a materials cycle offers a comprehensive framework for considering
directions of action on national materials issues, particularly as
they relate to materials science and engineering. In that context we
have encountered several recurring themes:

- Materials, energy, and the environment are parts of the
  same vast system; policies and programs that deal with
  one will falter unless they take full account of the
  other two on the same level and against the backdrop
  of the materials cycle.

- Materials science and engineering will play a pivotal
  role in managing and conserving this country's material,
  energy, and environmental resources, presenting as it
  does a total body of science and engineering that can be
  invoked in a sophisticated -- perhaps unprecedented --
  manner to help solve societal problems.

- Interdisciplinary research has become essential to progress
  in complex fields like materials, the environmental
  sciences, and medicine, but the universities generally
  harbor some resistance to interdisciplinarity going well
  beyond that needed to preserve the separate, and indispensable, scientific and engineering disciplines.

- Materials science and engineering displays an unusually
  close and continuous linkage between basic research and
  ultimate applications, together with a combination of

responsiveness and creativeness that holds strong potential
for upgrading technologies regarded as socially and
economically important

- Advances in materials and related fields feed on bodies
of knowledge that require steady replenishment by research
and development, suitably funded and carefully balanced
between the basic and the applied.

The 24 recommendations that follow, we believe, propose realistic actions consistent with these themes. The Recommendations fall
naturally into five groups, depending on the emphasis of the action
proposed: technical, governmental, industrial, academic, and professional. The sequence of the Recommendations should not be construed
as rank ordering in any sense.

Recommendations for Technical Action

Materials Research and Development Required for Progress in Energy Technology

The pressing demand for energy in this country is creating
problems that simply cannot be solved without skillful exploitation of
materials science and engineering. We must learn to generate, transmit,
store, and use energy more efficiently and within appropriate environmental constraints. All too often, it seems that the developers of
new technologies have counted heavily on the expectation that improved
materials would somehow be discovered as needed. This risk is too
great to take in the energy field. Inadequate materials hamper our
current fossil- and nuclear-fuel technologies; and lacking new or

sharply improved materials, some advanced energy systems may never be reduced to practice. Federal leadership is essential for the coordinated development of energy technologies, but industry, despite the pressures of short-term economic survival, can do much to help solve the related long-range materials problems.

IT IS RECOMMENDED THAT both government and industry define clearly, and ensure that close attention is being given to, those areas of materials research and development likely to be critical for significant progress in methods of generating, transmitting, storing, and using energy.

1

Energy

This recommendation should be implemented in the federal government by the highest-ranking office concerned with energy policy. In industry, action should be pressed by organizations like the new Electric Power Research Institute. Upgraded materials are required for high-temperature gas turbines, for breeder reactors, for magneto-hydrodynamic generators, for energy-storage devices, and for super-conductor technology. The unique advantages of solar energy warrant coordinated attack on the pertinent materials problems, with adequate long-term funding by the National Aeronautics and Space Administration and the National Science Foundation. Electric power from nuclear fusion is not a certainty, but pending a demonstration of technical feasibility, the presumed materials demands of the process should be studied critically to minimize the possibility that they may become the limiting factor. Such materials problems, together with those currently inhibiting the nuclear-fission technologies, should

receive sustained attention from the Atomic Energy Commission.
(Pages 70-76)

Materials Expertise in Environmental Management

A large fraction of man-made pollution results from activities
involving materials (even excluding foods and fossil fuels, as we do
here). It follows that we can solve many environmental problems by
moving materials through the materials cycle more carefully. The
consequent job for materials science and engineering -- meshed closely
in practice, with product design -- is to discover and develop
materials and processes that ease the pressures on the environment
without corresponding sacrifice in function and cost. This approach,
which cuts across an unusually wide range of disciplines, social as
well as technical, is invoked somewhat today, but hardly to the degree
that is possible and necessary. For the most part, the materials
community is not yet oriented toward the complex issues of environmental systems. Moreover, those concerned with such questions may not
have fully appreciated the potential of materials science and
engineering in these matters.

IT IS RECOMMENDED THAT the interdisciplinary capabilities of materials
science and engineering be applied more inten-
sively along a broad front on materials-related

2

Environment

environmental problems, with emphasis on the
materials cycle and its energy and environmental
subcycles.

This recommendation should be implemented by the Environmental Protection Agency, and should also be heeded by industry and the universities. Elements of a systems approach to environmental quality, in which strong participation of the materials community is vital, pertain not simply to products, but also to materials development, selection, and processing; discovery of substitute materials and functional alternatives; product design and manufacture; product/environment interaction; materials reclamation and disposal; and instrumentation for pollution monitoring and control. The Environmental Protection Agency might find unusual opportunities for pursuing such topics in existing federal laboratories, as in Recommendation 12. (Pages 12-13, 41, 43, 56-63, 86-89)

Materials Emphasis in Goal-Oriented Research

Potential scarcities of certain materials, the country's current shift in technological emphasis toward civilian-oriented goals, and recent trends in consumer and environmental legislation, all combine to raise unprecedented and challenging materials-related questions. The results of COSMAT's priority analysis show, among other things, that in some areas significant progress will occur only if materials research can surmount major roadblocks; in other areas, materials research can move us ahead markedly even when materials may not be limiting factors.

IT IS RECOMMENDED THAT organizations, including government, that support or perform goal-oriented research, especially in civilian-directed areas, ensure that work on end products is accompanied by adequately-supported research on related problems in materials including, where appropriate, special attention to the development of substitutes based on the more abundant materials.

3

Goal-Oriented
Research

The topics of highest priority in goal-oriented materials research established by our analysis are summarized according to areas of national impact in Table 14. (Pages 54-96)

Applied Materials Research of Broad Implication

The nation's civilian technologies and still-significant progress in defense and space depend for success on sustained, strong efforts in applied research on materials. It is critical that such research include work in certain materials areas of broad implication, namely, those spanning a range of missions or end uses, and so can be regarded as generic. Because this type of research often lies between basic and mission-directed research, it runs the risk of receiving inadequate support. In these areas -- such as corrosion, testing and characterization, and toxicity -- what is wanted is widely-applicable knowledge and methods, rather than one-shot empirical solutions to individual difficulties. It appears, therefore, that many of the problems we have in mind can be investigated fruitfully by scientists trained originally in basic research.

IT IS RECOMMENDED THAT investigators in the fundamental aspects of materials, particularly at universities, exercise initiative in identifying and pursuing opportunities in generic applied research on materials, and that they recognize the importance of such research in maintaining the vitality of materials research and development. (See Recommendation 10.)

4

Generic Applied Research

Priorities in generic applied materials research where specialized knowledge would certainly prove useful appear in Table 16. From the industrial standpoint, the action recommended could be advanced by cooperative funding of programs in universities, research institutes, and independent laboratories. For federal agencies that support materials research and development, the generic applied work discussed here is an essential element in establishing properly balanced programs. (Pages 97-102, 106-134)

Research on Fundamental Properties of Materials

The vitality of materials science and engineering also depends on sustained basic research to advance the fundamental understanding that allows the behavior of electrons, atoms, and molecules to be related to the world of product function and performance. It is essential that we add steadily to the reservoir of new basic knowledge on materials, a reservoir tc be tapped eventually in ways that cannot now be foreseen. Our predictive ability is relatively good for elemental and single-crystal materials, particularly those with potentially useful electronic properties, though many questions remain. For the multitude of more complex materials, however, we have made only the barest beginnings toward developing the necessary fundamental concepts.

IT IS RECOMMENDED THAT federal agencies and industries that perform or support basic research on materials phenomena encourage adequate attention to studies of relatively simple (model) solids while, at the same time, placing increasing emphasis on materials which are more complex in composition and structure.

5

Research on
Fundamental
Properties

This recommendation is directed primarily to the National Science Foundation and the mission-oriented federal agencies that support basic research. The focus proposed, however, is appropriate also for companies where management is receptive to the prospect of longer-term payoff. Promising topics for basic research include: interatomic forces, chemical bonding, and lattice stability; microscopic mechanisms of phase transitions; the amorphous, disordered state of matter; impurity and defect phenomena in solids; surfaces; one- and two-dimensional systems (e.g., linear molecules and interfaces, respectively); structure-property relationships in polymers; collective behavior of excited systems of atoms and electrons; and the dynamics of nonequilibrium systems. (Pages 97, 103-104, 134-153)

Renewable Resources as a Raw-Materials Base for Polymers

The tonnage of synthetic polymers -- plastics, fibers, and rubbers -- produced annually in the United States is now comparable to that of nonferrous metals. About 90 percent of the output is based on petroleum and natural-gas liquids; while polymers account for less than 5 percent of our consumption of these hydrocarbons, the implicit conflict with energy requirements is likely to intensify. Although oil shale and coal might be developed as raw-materials bases for polymers, it is nevertheless attractive to consider the technical feasibility of deriving synthetic polymers from renewable resources despite the fact that in the short range, hydrocarbons have a substantial economic edge.

IT IS RECOMMENDED THAT studies be undertaken on the feasibility of

6

Renewable
Resources

using renewable resources, including organic wastes, as a raw-materials base for synthetic polymers.

This recommendation should be implemented by the Board on Agriculture and Renewable Resources of the Commission on Natural Resources, National Research Council. The analysis we recommend would emphasize topics such as: cellulose from wood, plants, and organic wastes as a major raw-materials base; the properties of cellulose-derived polymers compared to those manufactured from hydrocarbons; and production of ethylene, the major monomer in synthetic polymers, from alcohol made by fermenting organic wastes. Each such topic, moreover, must be considered in terms of its relative ecological impact, including the biodegradability of the resulting synthetic polymers.   (Pages 57-58, 88-89)

Materials Selection and Product Design to Facilitate Recycling

The need to raise the recycle rates of many materials is likely to become more intense, for both economic and environmental reasons. Skillful integration of materials selection with product design can ease the dismantling and separation of components for recycling, but this approach is not always straightforward. Metals like those in a shredded automobile, for example, tend to be degraded with each recycle, although they may be suitable for functions less demanding than the original ones. The same is true of many other materials, including blended plastics, ceramics, composites, and glass.

IT IS RECOMMENDED THAT the resources of materials science and engineering be deliberately exploited and extended
to upgrade the recyclability of materials

7

Recyclability

through materials development and selection,
meshed carefully with product design, and
through the development of new recycling
processes.

This recommendation should be implemented by the Environmental Protection Agency. Typical targets for materials science and engineering would include the development of materials -- ceramic, metallic, and polymeric -- in which additives and alloying elements do not interface with the recycling of the base material. The eventual goal is to combine materials selection, product design and manufacture, and recycle processing in a systems approach to the optimization of new product development. (Pages 62-63, 87)

Recommendations for Governmental Action

Federal Policies and Programs in Materials

Demands for materials, old and new, cannot fail to intensify in the years ahead. It is imperative that the nation look more closely at how best to meet the changing requirements for materials, not only in terms of conventional market factors, but also in the light of consumer attitudes, environmental pressures, and international relations. Materials science and engineering, applied creatively to the materials cycle, can do much to integrate modern materials technology into federal policies on materials supply and usage and the interrelated policies on energy and the environment. We also see a major role for materials science and engineering in federal mechanisms for technology assessment. Such approaches can work well only when objectives are clearly delineated, and yet materials-related federal responsibilities today are diffused among many agencies and advisory bodies that seem to have no unifying goals. This fragmentation is particularly disadvantageous from the standpoint of developing effective national policies pertaining to energy, the environment, and materials.

IT IS RECOMMENDED THAT the federal government equip itself with analytical and advisory capabilities for addressing national materials policies and programs in the context of the materials cycle and the associated energy and environmental requirements, drawing deliberately on the knowledge and experience of the nation's technical community; and that each governmental agency involved with materials be made responsible, in its planning, functions, and technological verification of programs, for taking materials issues into full account on the same level as, and in concert with, energy and environmental issues.

8
Federal Materials Policies

This recommendation is in keeping with corresponding recommendations of the National Commission on Materials Policy; it relates also to the implementation, by the Department of the Interior, of the Mining and Minerals Policy Act of 1970. A significant purpose of the coherent approach to materials questions being proposed here is the stimulation of research and development in materials to improve the country's competitiveness in world markets and to optimize its consumption of resources. Additional points are: to give materials issues proper weight in national policies; to provide for an integrated approach to the entire materials cycle, from resource identification through processing, engineering applications, end use, disposal, and recyclability; and to provide leadership for international cooperation in the materials field. Critical to the implementation of this Recommendation is the ability to make predictions relative to the materials cycle. This ability relies in turn on an analytical capacity -- including data bases and econometric modeling -- whose development poses a strongly interdisciplinary challenge to both physical and social sciences. (Pages 4-6, 16-21, 28-29)

International Cooperation on Materials-Related Issues

Nations have long contended with common problems of supply and trade in materials. Other materials-related international activities include the International Standards Organization, the newly-formed Environmental Program of the United Nations, and the U.S.-U.S.S.R. program of cooperation in scientific research, in which catalytic materials are among the areas of interest. We see much to gain, however, from more extensive international cooperation on materials-related questions.

IT IS RECOMMENDED THAT the United States, through international bodies as well as bilaterally, press for greater international cooperation in such matters as setting materials standards, care of the environment, conservation of resources, materials research of wide-ranging import, and exchange of materials scientists and engineers (and specialists in other fields); and that private organizations

9

International
Cooperation

involved in materials seek more extensive
cooperation with their counterparts abroad.

Responsibility for implementing this recommendation would lie
primarily with the Departments of State and Commerce, although the
federal science advisory structure should play a prominent part.
Scientific and engineering societies should be urged to undertake new
initiatives to promote freer flow of professional and technical information around the world. The federal action recommended, among its
potential benefits, could help this country develop its materials
policies in cognizance of those of other nations. (Pages 13-15)

Federal Support for Basic and Applied Research in Materials

Basic research in materials must be balanced properly with
applied research if we are to maintain the close linkage and relatively
short time scale -- 10 to 20 years -- between basic research and
applications that has characterized materials science and engineering.
The traditional product- or mission-directed applied research is
best supported, in general, by industry or mission-oriented federal
agencies. There is in addition, however, considerable need for generic
applied research in materials (see Recommendation 4). It is this
broadly applicable research that should be balanced judiciously with
basic research to achieve well-rounded materials programs.

IT IS RECOMMENDED THAT those federal agencies responsible for funding

10

basic or applied research encourage the materials community to identify and attack problems

Federal Support for
Materials Research

in generic applied research, in line with

Recommendation 4.

This recommendation is directed primarily to the National Science Foundation. Funding of basic research in materials by the NSF Materials Research Division (and by other agencies) is well established. Support of generic applied research could be undertaken by that Division, but is particularly suitable for the Foundation's program in Research Applied to National Needs, given the necessary administrative and financial resources. Generic applied research on materials, where feasible, could also be undertaken to advantage by mission-oriented federal agencies. (Pages 16-21, 32-36, 97-104, 106-134)

Coordination of Materials Research within the Federal Government

The federal government conducts and supports extensive research and development in materials. Liaison among the agencies involved is handled by the Interagency Council for Materials and by other, less formal groups. These bodies as now constituted are unable to gather and analyze information to the degree required to optimize the allocation of the pertinent federal resources; they also lack the influence to motivate adequate federal response. More detailed liaison will be needed in the future to facilitate transfer of knowledge and to avoid unnecessary duplication of effort while identifying new opportunities in materials research and development.

IT IS RECOMMENDED THAT an appropriate federal body be assigned the authority to review regularly the allocation of federal funds for materials research and development, to assess the progress of such research and development and to recommend changes in emphasis in terms of national objectives, such changes to be implemented through the Office of Management and Budget.

11

Coordination of
Federal Materials
R&D

The action proposed here is a necessary part of the analytical and advisory capability called for by Recommendation 8. The reviews, assessments, and recommendations concerning materials research and development should go annually to the relevant agencies and to the federal science advisory structure. The federal body named in this recommendation should also move to codify and make widely accessible the extensive technical information on materials generated in governmental programs. (Pages 16-21, 32-36)

Effective Use of Federal Laboratories

Federal laboratories, including federally-funded research and development centers, perform more than half the applied and basic research in materials supported by the federal government, or roughly twice the average for federal research spending in all fields. These laboratories have worked mainly in materials-limited areas in defense, nuclear energy, and space. As the civilian-oriented agencies begin to integrate materials science and engineering into their programs, in response to shifting national priorities, they should find it advantageous and economical to tap the large existing federal resource in materials research and development.

IT IS RECOMMENDED THAT civilian-oriented and other governmental agencies take full advantage of existing federal facilities and personnel to harness materials science and engineering to emerging programs, and that this federal resource in materials be utilized both in a consultative capacity and in performing the indicated research and development.

12

Use of Federal
Laboratories

The agencies themselves should be primarily responsible for implementing this Recommendation. Information and coordination should be provided by whatever federal body is designated in the implementation of Recommendation 11. The action recommended, among its other benefits, would help avoid costly losses in technological momentum that can result from discontinuities in federal programs. (Pages 8-21, 32-36)

Recommendations for Industrial Action

Integration of Materials Science and
Engineering with Design and Manufacture

Progress in experience-intensive or low-technology industries has been limited by relatively low investment in research and development, but it is also true that such industries often have not sufficiently exploited existing knowledge in materials science and engineering. In product development, for example, it is highly desirable and sometimes critical for materials specialists to work closely from the start, and on an equal footing, with design and production engineers. This collaboration can lead to more economical design, fewer startup problems in manufacturing processes, and improved product performance and reliability. Indeed, the alliance of materials specialists with design and production experts will grow ever more crucial as materials operations react to the mounting pressures of consumer, energy, and environmental requirements.

IT IS RECOMMENDED THAT technical management in industry make strong efforts to integrate materials science and engineering with product design and manufacture, as employed most effectively in the science-intensive or high technologies of aerospace, electronics, and nuclear energy.

13

Integration of
Materials Knowledge
With Design and
Production

This recommendation calls for bold industrial initiative, which could be encouraged by the Industrial Research Institute but in the end must be spearheaded by technical management. The experience-intensive industries will find no lack of opportunities for progress -- at a minimum the development of a keener awareness of the limitations of materials and of the capabilities of process and quality control. It has been estimated, for example, that perhaps half the corrosion in this country could be avoided by applying the available materials knowledge to industrial and consumer products. (Pages 46-63)

Stimulation of Materials Science and Engineering in Civilian Technologies

The universality of materials suggests that the stimulation of industrial programs in materials science and engineering can be a powerful stimulus to civilian technologies in general. In some materials-intensive industries -- construction, housing, materials fabrication -- many companies are relatively small and unable to develop or readily adopt new technology. Even large companies may be unable to justify research and development because the risks are too great, the markets too small, or both. Federal agencies that traditionally have provided both research funds and markets for high technologies like aerospace have no obvious counterparts among the civilian-oriented agencies. Such problems can be eased and sometimes solved by measures such as tax incentives and procurement regulations, but we propose that enterprising action in materials science and engineering warrants serious attention.

IT IS RECOMMENDED THAT fragmented industries be encouraged to conduct cooperative research and development in materials related to their product needs; that, in civilian-oriented technologies where extensive federal procurement can be anticipated or, alternatively, where markets are small, the pertinent federal agencies support industrial R&D on materials through the phases that entail unacceptable risk or long lead times as judged by realistic commercial practice; and that the federal government and trade associations jointly stimulate, on an experimental basis, the establishment of a small number of national or regional programs in broadly applicable, product-related materials science and engineering.

14

Materials Research
in
Civilian
Technologies

The Department of Commerce should take the lead in implementing this recommendation. Active participation would be required by the Department of Justice, the National Science Foundation, the federal body named under Recommendation 11, the mission-oriented federal agencies, trade associations, and individual companies. State governments should also seek ways to participate. The cooperative R&D might involve joint support of an industrial materials research center or of materials research programs in company, university, or independent laboratories. Civilian-oriented areas appropriate for federal funding of industrial research include construction and biomedical materials. The national or regional programs we propose would concentrate on generic problems like materials shaping, materials joining, and friction and wear (see Recommendation 4). These programs should be established preferably where suitable expertise already exists, particularly at universities and in federal laboratories. An example of what might be done is the longstanding cooperative program of the National Bureau of Standards and the American Dental Association in developing advanced dental materials. The actions proposed in this Recommendation would complement existing technology-incentives experiments in the Department of Commerce and the National Science Foundation. (Pages 3, 14, 51-53)

Research Needs in Bulk-Materials Industries

Bulk-material industries tend to invest less heavily in research than does industry generally, although they appear not to lack diverse technical challenges. With little ability to generate basic knowledge themselves, such mature industries can become incapable of evaluating and using basic knowledge generated elsewhere. Companies in these industries, if they do not have even small cadres of skilled scientists, performing comparably to and communicating with their academic peers, may find themselves literally unable to solve technical problems of clear commercial import. The danger seems to have been recognized in Japan, for example, where the recent trend in such industries has been to invest increasingly in research. In the United States it is now urgent for the materials-producing and processing industries to begin or enlarge research programs aimed at greater efficiency in processing

and manufacturing, particularly in the face of mounting ecological
pressures.

IT IS RECOMMENDED THAT corporate managements in bulk-materials industries make sure that the nature and scope of
their research programs, especially with regard
15
to materials processing and manufacturing
Research in
Bulk-Materials
Industries
methods, are such that new knowledge and
techniques generated elsewhere can be effectively
assimilated.

This recommendation calls in some cases for thorough rethinking
of industrial practice in research, with a view to maintaining at least
the ability to evaluate and use new developments discovered elsewhere.
A prime opportunity appears to lie in materials extraction, processing,
and recycling, where entirely new technologies may be required to
obtain useful products from huge tonnages of very low-grade and widely
disseminated deposits at acceptable energy and environmental costs.
Also required are more competitive manufacturing processes, involving
especially the continuous production of metal parts or shapes from the
fluid state (liquids or powders). These challenges in extraction,
processing, and manufacturing call for new automation techniques
involving servomechanisms, minicomputers, and materials-critical
sensors based on the interactions of diverse forms of matter with
acoustic, electromagnetic, and other forms of radiation. All of these
directions suggest, on the whole, a pressing need in mature industries
for greater participation by scientists and engineers who may not be

traditionally associated with those industries, such as professionals in computers, electronics, lasers, and nuclear reactors. (Pages 51-53)

Value of Specialized Research Centers

A major contributor to the achievements of materials science and engineering during the past two decades has been the revolution in research equipment, instrumentation, and analytical tools. The current requirements of the field for such equipment entail relatively modest cost in the scale of modern science and technology, but these requirements are not being met in some areas, for example, in flammability, nondestructive testing, robotics, and biomaterial evaluation. Central facilities can be an economical means of satisfying such needs.

IT IS RECOMMENDED THAT a small number of specialized regional and national centers be established cooperatively by industry, the universities, and government

16

Specialized Research Centers

for providing research and equipment services in materials science and engineering on a broad basis, and that these facilities be centered on existing capabilities, intellectual as well as physical, that are already of high quality and that can be made readily accessible to the technical community.

Industry should take the lead in implementing this recommendation and should encourage the universities to make appropriate facilities available for the purpose. The centers should aim to become largely self-supporting in service work. The staffs should also do research, however, in order to be able to provide a well-rounded capability, and this would probably require sustained outside support. The department of Commerce and the National Science Foundation might participate in this work, in part because of their current experiments in stimulating civilian technologies. Equipment for materials research includes high-voltage electron microscopes, nuclear reactors, and particle accelerators, requirements that can probably be met with existing federal and university facilities, providing they are funded adequately and made widely accessible. Among the needs in materials engineering are programs on flammability and on nondestructive testing. Collaboration in such areas would be natural for industrial and nonprofit laboratories. A central facility is warranted to develop methods of evaluating biomedical materials and related standards; the National Institute of Health and the National Bureau of Standards might jointly establish such a unit. A center is also desirable for research on the automation of industrial processes, which would require close interaction among materials scientists and engineers and specialists in information processing (electronics and computers). (Pages 28-30, 86, 90,93-96, 131-134)

Recommendations for University Action

Need for Interdisciplinary Programs in Universities

To solve technological problems as well as to advance science often calls for interdisciplinary attack, and the materials field offers useful lessons in this respect. Yet the practice of interdisciplinary research and education at universities, including materials research centers, is impeded by the disciplinary and administrative characteristics of the institutions themselves. It is inhibited also by the internal structure of some of the main research-supporting agencies, including the lack of balance with respect to disciplines and materials in the staffing of those agencies.

17

Interdisciplinary
Activities

IT IS RECOMMENDED THAT universities intensify their efforts to build interdisciplinary activities in research and education; that the barriers to interdisciplinarity in universities be examined critically; and that guidelines be developed for recognizing and rewarding academic achievement in interdisciplinary and interdepartmental programs.

This recommendation must be implemented by the universities themselves, although the American Council on Education could also undertake a study of the difficulties encountered by interdisciplinary programs. Materials science and engineering is one of several logical vehicles for such an effort. Federal agencies can encourage interdisciplinary work at universities through appropriate incentives and support, not only for research, but also for training students in the interdisciplinary approach to problem-solving. Supporting agencies should recognize, however, that suitably strong programs must be maintained in the traditional disciplines, which are essential to sound interdisciplinary activities. (Pages 23-27, 37-41)

Materials Education for Physical Scientists and Engineers

More than a decade ago, a comprehensive report[*] on engineering education pointed out the importance of education in materials for all engineering undergraduates. The makeup of the nation's manpower in materials science and engineering including as it does large numbers of engineers, physicists, and chemists, as well as holders of materials-designated degrees, reinforces this view and extends it beyond engineering students in the physical sciences.

IT IS RECOMMENDED THAT undergraduate education in the physical sciences as well as in engineering provide opportunities for a flexible content of solid-state topics relevant to materials science and engineering.

18

Materials Education
for Undergraduates

This recommendation invites attention by the National Science Foundation and the National Institute of Education, as well as by the academic community. The exposure we propose might also consist of

---

[*] Grinter, L. D., "Report on Evaluation of Engineering Education (1952-55)," _Journal of Engineering Education_, 46, 25 (1955)

elective subjects, minor programs, or double majors, depending on the field and level. The concept of structure/property relationships could be emphasized in certain physics, chemistry, and engineering subjects. (Pages 28-33)

Balance in Materials-Degree Programs

University departments offering materials-designated degrees have, in the main, built into their curricula a suitable scientific base in physics, chemistry, and the pertinent engineering sciences. Substantial imbalances exist, however, in other areas important to the long-range effectiveness of materials science and engineering.

IT IS RECOMMENDED THAT, depending on local circumstances, materials-degree programs provide increased emphasis on such engineering topics as: materials preparation and processing; polymer technology; design and systems analysis; computer modeling; relations among the properties, function, and performance of materials; and that research in these areas be included.

19

Curricular
Balance

The academic community should implement this recommendation. We believe that the curricular balance proposed will improve the education of a large fraction of the materials graduates who will pursue careers outside the university. (Pages 39, 54-96)

Block Funding of Materials Research Centers

COSMAT's inquiries into the existing materials research centers at universities confirm that the federal experiment of block funding, with research projects and facilities selected and managed locally, is a sound means of encouraging research of high quality. Performance at individual block-funded institutions has been uneven, however. It appears that some focusing of the associated research is usually desirable if strong interdisciplinary activities are to develop. And although central facilities have shown their potential for increasing the sophistication and output of materials research, actual working interactions in cooperative research on a given campus appear to depend more on local leadership by faculty and administration.

IT IS RECOMMENDED THAT support of materials research centers through block grants be accepted as an established funding method; that block grants be awarded and renewed on a competitive basis and provide for forward or step funding; that, in addition to support for individual scientists, some concentration of effort be encouraged to take advantage of local research specializations; and that appropriate parts of the center programs be oriented toward materials systems (integrated combinations of materials), processing, and applications.

20

Materials Research
Centers

This recommendation applies primarily to the National Science Foundation, the Atomic Energy Commission, and the National Aeronautics

and Space Administration. Of the federal budget for university
research in materials, the proportion directed to materials research
centers seems generally adequate to retain overall quality and flexibility; step funding will lessen problems caused by federal program
changes or budget reductions. Focused efforts would offer a promising
opportunity for civilian-oriented federal agencies to stimulate pertinent materials research at universities by contributing to the support
of block-funded programs. (Pages 37-40)

Recommendations for Professional Action

Roles for National Advisory Groups

Numerous groups advise parts of the federal government on
special aspects of materials, but the two with continuity and wide
scope are the National Materials Advisory Board and the Committee on
Solid State Sciences, both within the National Research Council of the
National Academy of Sciences and the National Academy of Engineering.
Each committee has dealt only with specific sectors in the field of
materials science and engineering, and neither has discretionary funds
with which to conduct studies.

IT IS RECOMMENDED THAT the National Research Council coordinate more
fully and draw effectively on the materials
interests and expertise available to it through
the two Academies in order to strengthen its
advisory capacity across the full spectrum of

21

Materials Advisory
Groups

materials topics, particularly where national
policies or goals are at issue.

As part of the action recommended, the National Materials
Advisory Board should continue to broaden its membership and materials
coverage so as to serve a wider range of industries and governmental
agencies. In addition, the Board and the Committee on Solid State
Sciences should be recognized more fully as complementary bodies and
utilized accordingly. Because of the recurring need to identify
national materials problems and opportunities, we expect that the
Board and the Committee between them will become an important source
of information and support for the newly established National Research
Council Commissions on Societal Technologies, Natural Resources, and
Peace and National Security. (Page 16)

Coordination of Activities by Professional Societies

Professionals in materials science and engineering are served
by about 35 technical societies. Until recently there has been no
mechanism to minimize overlaps in programming and otherwise coordinate
the interests of materials professionals, many of whom must belong to
several societies to cover their professional and technical needs.
Formation of the Federation of Materials Societies in 1972, was a
major progressive step; of the 17 broadly based societies invited to
participate, nine had joined by October 1973.

IT IS RECOMMENDED THAT professional societies concerned with

22

materials coordinate their programming and

information-distribution functions, and that

Coordination of
Professional
Societies

the societies actively support and participate

in the Federation of Materials Societies.

The Federation is a very promising mechanism for achieving a framework within which professionals in materials science and engineering will be able to recognize themselves as members of the field as a whole. Such cohesion will help attract well-qualified entrants to the field and will help ensure the proper allocation of resources to it by government and industry. The Federation in turn should offer its services to appropriate public and private bodies wherever it can be useful in matters involving materials. The Federation should also facilitate efforts among the societies to organize their work in technical programming, publications, and information-retrieval systems. A well-coordinated program is likewise required to increase public awareness of the underlying importance of materials in achieving national goals and of the role of materials science and engineering in securing the benefits of materials to mankind. (Pages 1-2, 23-33)

Greater Flexibility of Materials Manpower

Government, industry, and the universities interact in various ways that tend to increase the technical flexibility of materials (and other) scientists and engineers. Examples include joint academic-industrial appointments and staff rotation, joint research projects, the Ford Foundation's one-year industrial residency program, the Commonwealth of Pennsylvania Resident Industrial Scholarships for short-term appointments, and the Research Associates Program of the National Bureau of Standards. The extent of such interaction on a national scale, however, is not commensurate with its potential value.

IT IS RECOMMENDED THAT government, industry, and the universities pursue arrangements ranging from temporary exchanges in personnel to joint academic-industrial appointments in order to promote greater interaction and flexibility among materials scientists and engineers from the various sectors.

23

Flexibility of Manpower

This recommendation could be implemented cooperatively by the Industrial Research Institute and the National Science Foundation. Precedents exist for the arrangements recommended, and they should be exploited in the materials field. Industry and government, for example, might look to the universities to become foci for national or regional pilot research programs and for specialized knowledge in materials science and engineering. The steps recommended also could serve usefully for state and local projects dealing with regional industries, technological planning, mass transit, special energy requirements, and environmental problems. (Pages 23-33)

Improved Statistics on Manpower and Funding

Serious shortcomings exist in the means for gathering statistics nationally on scientific and engineering manpower, employment, and associated resources. The data assembled by the Office of Education (Department of Health, Education, and Welfare) on degrees awarded annually are not coordinated with those in engineering collected by the Engineering Manpower Commission. Both sets of data are inadequate for analyses of materials-designated and related degrees. The National Science Foundation's National Register of Scientific and Technical Personnel, which provided important data on manpower characteristics, has been discontinued. The NSF data on funding for education and for research and development are at a level of detail that limits their utility for long-range planning. The federal research-funding data gathered by the Interagency Council for Materials are likewise incompletely developed.

IT IS RECOMMENDED THAT the National Academy of Sciences and the National Academy of Engineering, with support from the National Science Foundation, reassess the national data-gathering mechanisms for manpower, employment, and funding in science and engineering and that they recommend to the Foundation the actions required to create an internally consistent system suitable for long-range planning on a disciplinary or multi-disciplinary basis.

24

Manpower and
Funding
Statistics

Implementation of this recommendation should be initiated by the National Science Foundation and coordinated with the Bureau of Labor Statistics (Department of Labor), the Office of Education, scientific and engineering societies, and other relevant groups. It is most important that the pertinent data be collected and organized in a form useful for analysis and planning in multidisciplinary areas such as materials science and engineering and the environmental sciences. A sound data base of the kind recommended is essential for effective federal planning and budgeting in the sciences, education, employment, and related areas. (Pages 30-41)

APPENDIX A

PRIORITIES IN MATERIALS RESEARCH

COSMAT QUESTIONNAIRE METHODOLOGY AND SOME RESULTS

The goal of the questionnaire used by COSMAT was to determine priorities among topics in basic and applied research in materials science and engineering, as viewed by scientists and engineers knowledgeable in the field. Some 2,800 questionnaires were mailed out using a mailing list which was selected to provide representative coverage of materials science and engineering. 555 of the responses were sufficiently complete to be included in the analysis.

Characteristics of Responding Group

Age: 50 and up, 262; 40-49, 214; 30-39, 74; under 30, 5

Highest Degrees: PhD., 379; Master, 78; Bachelor, 62; Blanks, 36

Discipline of Highest Degree: Metallurgy and Ceramics, 172; Physics, 153; Chemistry, 95; Engineering, 71; Other, 8; Blanks, 56

Employer: Industrial, 215; Academic, 187; Government, 120; Non Profit, 16; Other, 17

Activity: Research, 350; Teaching, 181; Development or Engineering, 122; Technical Management, 262; General Management, 76; Other, 52

Number Managed (if a manager): over 100, 81; 10-100, 163; less than 10, 80

Rating Process

Respondents rated priorities for basic and applied research in various specialties of materials science and engineering on a five-level scale. Rating numbers were then calculated from:

$$\frac{100\ (1)\ +\ 75\ (2)\ +\ 50\ (3)\ +\ 25\ (4)}{(1)\ +\ (2)\ +\ (3)\ +\ (4)\ +\ (5)}$$

where (1) is the number of "very high" responses, (2) is the number of "high" responses, etc. This gives a rating number for each specialty between 0 and 100, where 0 would mean all "very low" responses and 100 would mean all "very high" responses. In the case of applied research, the priority rating for each specialty was obtained on the basis of its relative importance to various national inpact areas and subareas. The respondents also rated their own familiarity with each specialty on a five-level scale, and a rating number for such familiarity was similarly calculated. The questionnaire covered 46 specialties in materials science and engineering, nine national impact areas, and 52 subareas.

The data presented here are condensed from a more extensive analysis of the replies to the questionnaire, which will be included in a later COSMAT report. The fuller report will also give break-downs of responses by various subgroups selected according to academic discipline, highest degree, age, type of institution, type of activity, and management level. These subgroups showed some individual differences, but by and large the responses of the various subgroups were remarkably similar.

Overall Importance of Materials Science and Engineering

Materials science and engineering can have different levels of impact on the various areas of technology in which materials are involved. To assess these differences, the questionnaire made use of nine Areas of Impact. The respondents were first asked to rate the overall importance of materials science and engineering to each of these Areas. Based on the average response, the Areas of Impact can be divided into three groups:

Very High Importance	Communications, Computers, and Control Defense and Space Energy
High Importance	Transportation Equipment Health Services Environmental Quality Housing and Other Construction
Moderate Importance	Production Equipment Consumer Goods

A similar analysis was also based on the responses of only those deemed to be particularly knowledgeable in the given Areas of Impact. Respondents who chose to rate a particular Areas or Subarea of Impact in detail were grouped together (by Area of Impact) and the responses in each of these groups was averaged (respondents could be in more than one group). This method of analysis provided a rating of the overall importance of materials science and engineering to each Area of Impact as rated by persons expert in it. The results classified the Areas of Impact into almost exactly the same rank order as shown above.

We conclude that, in the assessment of the general importance of materials science and engineering to the various Areas of Impact, opinions of the "experts" and the overall opinions matched fairly closely.

Methodology of Data Handling

A list of specialties within materials science and engineering was presented in the questionnaire, divided into three categories: Properties of Materials, Classes of Materials, and Processes for Materials. These are listed in Table A-1. Each respondent was asked to indicate his level of familiarity with each of these specialties and to rate the priority for Basic Research (research not specifically identified with any one Area of Impact) for each specialty. The familiarity and priority responses for each specialty were arithmetically averaged over all the respondents and the results are presented in Table A-1.

To assess the priorities for Applied Research, the Area of Impact is important. Within each Area of Impact, several Subareas were identified. These are listed in Table A-2. Each respondent was requested to select up to five Subareas of Impact and, for each, to rate the importance of Applied Research and Engineering in each of the specialties under Properties, Materials, and Processes. In addition, for each chosen Subarea of Impact, priority ratings were obtained for research activities according to the various disciplines comprising materials science and engineering. The respondents also indicated their degree of familiarity with each of the disciplines.

TABLE A-1

Priority Ratings for Basic Research in Materials Science and Engineering,
Arranged according to Specialties

SPECIALTY	Familiarity of Respondents	Priority for Basic Research
Properties of Materials		
Atomic Structure	61	68
Microstructure (Electron Microscope Level)	54	69
Microstructure (Optical Microscope Level)	61	53
Thermodynamic	60	64
Thermal	54	57
Mechanical and Acoustic	60	70
Optical	48	61
Electrical	55	66
Magnetic	45	52
Dielectric	43	52
Nuclear	41	60
Chemical and Electrochemical	49	70
Biological	20	56
Classes of Materials		
Ceramics	54	72
Glasses and Amorphous Materials	52	68
Elemental and Compound Semiconductors	47	62
Inorganic, Nonmetallic Elements and Compounds	50	59
Ferrous Metals and Alloys	58	59
Nonferrous Structural Metals and Alloys	53	63
Nonferrous Conducting Metals and Alloys	51	57
Plastics	40	65
Fibers and Textiles	28	46
Rubbers	24	42
Composites	45	70
Organic and Organo Metallic Compounds	28	51
Thin Films	43	62
Adhesives, Coatings, Finishes, Seals	33	58
Lubricants, Oils, Solvents, Cleansers	23	43
Prosthetic and Medical Materials	21	54
Plain and Reinforced Concrete	21	31
Asphaltic and Bituminous Materials	16	27
Wood and Paper	20	30
Processes for Materials		
Extraction, Purification, Refining	43	60
Synthesis and Polymerization	33	61
Solidification and Crystal Growth	59	66
Metal Deformation and Processing	49	56
Plastics Extrusion and Molding	29	43
Heat Treatment	58	55
Material Removal	44	51
Joining	47	61
Powder Processing	43	56
Vapor and Electrodeposition, Epitaxy	43	58
Radiation Treatment	35	55
Plating and Coating	42	55
Chemical	39	51
Testing and Nondestructive Testing	62	71

TABLE A-2

Responses Received, Arranged according to
Areas and Subareas of Impact

Code Number	Areas and Subareas	Number of Responses
10	COMMUNICATIONS, COMPUTERS, AND CONTROL	31
11	Commercial Radio and TV Equipment	10
12	Computers	66
13	Electronic Components	144
14	Equipment for Guidance and Control of Transportation	8
15	Teaching Equipment	14
16	Telephone and Data Networks and Equipment	41
	Total	314
20	CONSUMER GOODS	10
21	Apparel and Textiles	20
22	Furniture	6
23	Household Appliances - Electronic (TV, Radio, hi-fi, etc.)	23
24	Household Appliances - Nonelectronic (refrigerators, ranges, air conditioners, vacuum cleaners, etc.)	19
25	Leisure and Sports Equipment	4
26	Packaging and Containers	34
27	Printing and Photography	25
	Total	141
30	DEFENSE AND SPACE	39
31	Military Aircraft	81
32	Missiles	38
33	Naval Vessels	25
34	Ordnance and Weapons	38
35	Radar and Military Communications	46
36	Spacecraft	54
37	Undersea Equipment	35
	Total	356
40	ENERGY	35
41	Batteries and Fuel Cells	100
42	Direct Conversion	62
43	Electronic Transmission and Distribution	64
44	Fuel Transmission and Distribution	9
45	Nuclear Reactors	92
46	Thermonuclear Fusion	54
47	Turbines and Generators	66
	Total	482
50	ENVIRONMENTAL QUALITY	28
51	Mining and Raw Materials Extraction	65
52	Pollution	83
53	Recycling and Solid Waste Disposal	94
54	Reliability, Safety, Maintainability	25
55	Substitution Opportunities	19
56	Working Conditions	10
	Total	324

TABLE A-2 (Cont'd.)

Code Number	Areas and Subareas		Number of Responses
60	HEALTH SERVICES		14
61	Artificial Organs		39
62	Medical Electronics		13
63	Medical Equipment (including dental)		10
64	Prosthetic Devices (including dental)		64
		Total	140
70	HOUSING AND OTHER CONSTRUCTION		21
71	Construction Machinery		1
72	Highways, Bridges, Airports, etc.		19
73	Individual and Multiple Unit Dwellings		44
74	Industrial and Commercial Structures		12
75	Mobile Homes		13
76	Plumbing, Heating, Electrical, etc.		20
		Total	130
80	PRODUCTION EQUIPMENT		6
81	Farm and Construction Machinery		10
82	Industrial Drives, Motors, and Controls		9
83	Industrial Instrumentation		15
84	Machine Tools		22
85	Process Equipment		43
		Total	105
90	TRANSPORTATION EQUIPMENT		23
91	Aircraft		48
92	Automotive		75
93	Guided Ground Transportation (rail, nonrail)		30
94	Water		4
		Total	180

Basic Research

The numbers in Table A-1 show a trend: generally speaking, the greater the average familiarity, the greater the average priority given. This can be seen graphically in Figure A-1, where the Priority Rating for Basic Research is plotted against the Familiarity Rating for each Property, Material, and Process specialty. In an attempt to take account of these interplays, relative priority levels were determined from the rating numbers by three different methods:

Uncorrected for Familiarity

Respondents were divided into four groups according to the discipline of their highest degree -- chemists, physicists, metallurgists (including ceramists), and engineers. The simple rank orders in which each of these groups placed the Property, Class, and Process specialties were determined. The four disciplinary groups were then given equal weight in arriving at average rating numbers for given specialties.

Corrected for Familiarity

Here an attempt was made to correct the rating numbers for the degree of familiarity. Priority/familiarity trend lines were established graphically for each specialty, and the rank orders of the specialties were determined as the trend line was swept through the plots. This was done for each of the four disciplinary groups, and again the groups were given equal weight in determining average rank orders.

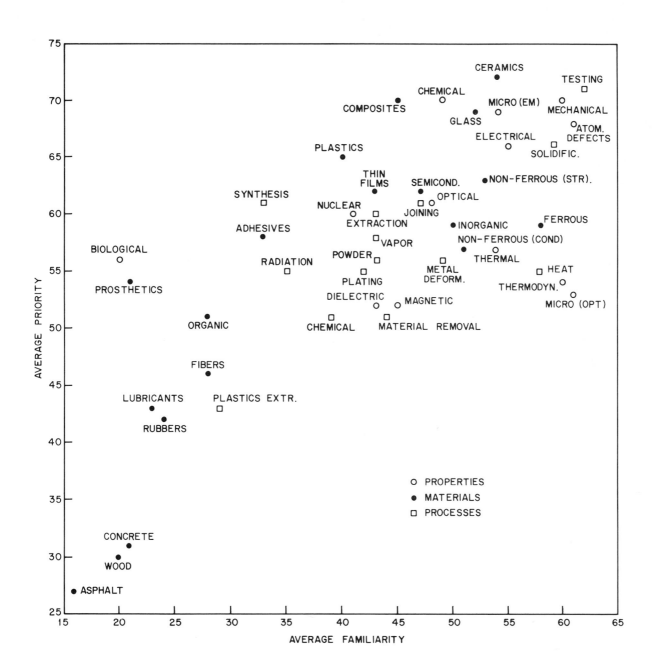

Figure A-1. Relationship between Priority Ratings for
Basic Research in Various Specialties and
Familiarity Ratings of the Respondents in
the Specialties.

Experts

Here we based the rank orderings on the opinions of the experts in each specialty. As previously mentioned, the experts were chosen by selecting those who indicated very high familiarity with the specialty. The responses of each group of experts (chemists, physicists, metallurgists and ceramists, and engineers) were then normalized so that the average response of each group over all specialties was the same. After this normalization, which was designed to give each group equal standing, despite their different numbers among the respondents, the various specialties were ranked according to the opinions of the experts in that specialty.

Table 17 in this report shows, on the left, the rank ordering for basic research in the various specialties, corrected for familiarity in the specialties. These ratings were converted to a four-symbol scale, where xxx designates very high priority, xx high priority, x moderate priority, and a blank indicates low priority. These indicators are listed in the second column on the right of Table 17. The uncorrected data were analyzed in the same way, with the results shown in the first column on the right. The rank ordering by experts in each specialty is shown similarly in the third column. The relative priority levels for basic research in the specialties depended somewhat on the method of analysis. For example, among Processes, research in radiation treatment was rated as low priority by the method uncorrected for familiarity, but was rated as moderate priority after correcting for familiarity, and as very high priority by the experts in radiation treatment. It was felt that particular significance should be attached to those cases in which the specialty was rated as very high priority both by the familiarity-corrected method and by the experts in that specialty. Such weighting is incorporated in the Overall Ratings listed in the fourth column on the right of Table 17.

Comments were requested in the questionnaire on specific research topics that the respondents considered important for each specialty. These comments are summarized below for the top-priority specialties. Here the asterisk denotes topics that were mentioned very frequently.

Properties of Materials

Chemical and Electrochemical. *Corrosion, stress corrosion, and oxidation (in aqueous systems, biological media, and hot gases; of aluminum, titanium, iron and steel, ceramics, thin films, concrete, and refractories; role of surface states, defects, and impurities)

*Catalysis (role of surface structure, impurities, free radicals, surface states and charges; nature of adsorption mechanisms)

Flammability

Electrochemical reactions

Chemical stability

Fundamental physics and chemistry of surfaces.

Biological. *Biodegradability (bacterial corrosion mechanisms, role of fungi, enzymes, hyphae, etc., fundamental mechanisms of interaction of materials and the environment)

*Biocompatibility (interaction between materials and blood
and tissue, immunological response to implants, protein interaction
with surfaces)

*Toxicity (ecological impact of materials, pollution standards,
mechanisms of heavy metal incorporation into biological compounds)

Classes of Materials

Ceramics. *Mechanical properties (tensile and impact strength,
toughness, ductility, creep, thermal shock resistance; effect of
flaws, effect of grain boundaries and microstructure)

Impurity effects (on diffusion, thermal, electronic, and
ionic conductivity; on magnetic and optical properties)

Plastics. *Durability (at high temperatures; degradation mechanisms)
Mechanical properties (relation to structure, bonding, side-chains,
cross-linking; role of thermal and mechanical history)

Composites. *Interface bonding properties (fundamentals of fiber-
matrix interface, compatibility and stability, stress transfer,
characterization on microelasticity scale, effect of molecular
variables in adhesives)

Mechanical properties (strength, ductility, fracture;
rheological properties; direction properties)

Prosthetic and Medical Materials. *Biocompatibility (materials with
physical and chemical properties matching adjacent hard and soft
tissue; nature of surface mechanisms of interaction of materials
with cells and proteins, blood adsorption; correlation between

in vivo and in vitro behavior; biorejection chemistry; electrical interaction with body fluids; durability)

New biomaterials (specific membranes, biological adhesives, glassy carbon, fluoropolymers, block polymers with ionic domains for controlled transport of long-term drugs)

Materials Processes

Testing and Nondestructive Testing. *Flaw detection (techniques for giving geometric description and location of flaws; crystallinity, texture; crack propagation, fatigue, creep; joint integrity)

Automatic monitoring (simultaneous checking of several parameters to monitor manufacturing processes)

Prediction of service life (accelerated aging testing, service environment testing, in-service indicators of incipient failure)

Exploitation of new physical phenomena and insights concerning interaction of radiation with matter

Techniques for testing special materials (biomaterials, nuclear materials, electronics materials, etc.)

Low-Priority Areas

The specialties rated as low priority for basic research (see Table 17) are of two general types. Some are specialties which have been heavily studied in the past, leading to diminishing returns for such research today. Possible examples are Ferrous Metals and Alloys and Nonferrous Conducting Metals and Alloys. Others are areas which have not been subjected to intensive basic research, such as Concrete, Asphalt, and Wood. In these cases our fundamental understanding may not have advanced to the point where research opportunities are clearly discerned, even by experts in the field.

Applied Research and Engineering

The responses for Applied Research and Engineering were treated as for the Basic Research, except that, since the respondents claimed to be knowledgeable in the Areas of Impact they selected, the overall averages for each specialty were used, rather than dividing the responses into the four groups according to disciplines. For the Areas of Impact, including all Subareas, the uncorrected rankings, ranking corrected for familiarity, and the rankings by experts were averaged (giving more weight to the latter two ratings), in order to arrive at the Overall Rating for each specialty relative to each Area of Impact, as indicated in Table 15.

Several specialties stand out with high-priority ratings almost across the board:

Chemical properties, for example, are rated as high priority for basic research and for several impact areas. From the comments it is clear that this assessment is related in part to the pervasive problems of corrosion, oxidation, and degradation, and the limitations they set on materials applications.

Mechanical properties also receive broad priority, as stronger and tougher materials are needed in nearly all fields of technology.

Of the materials classes, plastics received the highest overall priority rating, reflecting the rapidly-growing use of these materials in a wide range of applications. Table 15 also indicates the broad importance of composite materials, non-ferrous structural metals and alloys, ceramics and adhesives, coatings, finishes and seals. Under processes, testing was of the most widespread priority, with joining, polymer synthesis, and plastics extrusion and molding also rated high in many areas.

Although the above specialties received the broadest priority ratings, in certain Areas of Impact other specialties were ranked of equal or greater importance. Biological properties, for example, received high ratings in the Environmental and Health areas. Semiconductors, glasses, prosthetic materials, and lubricants ranked high for specific impact areas, as did the processes of vapor deposition and chemical processing.

It is obvious that the selected impact areas are very broad in scope. As a result, some specialties which rated low in particular impact areas were found to have high ratings in certain subareas. For instance, Electrical Properties were accorded only moderate priority in the area of Energy, but high ratings in the Subareas of Batteries and Fuel Cells, Direct Conversion, and Electrical Transmission and Distribution, and low ratings in the Subareas of Nuclear Reactors, Thermonuclear Fusion, and Turbines and Generators.

The written comments of the respondents relating to needs in Applied Research and Engineering are summarized below. Only the comments on the specialties rated as "very high priority" or "high priority" for Applied Research and Engineering in the various Areas of Impact are included here. Three asterisks indicate very high priority and two asterisks indicate high priority.

Communications, Computers, and Control

Properties

***Electrical: memories; solid state circuitry, large scale integration, display devices, Josephson devices, charge-coupled devices; miniaturization; reliability

**Atomic Structure: perfection; quality of crystals; surface effects; electromigration; ion implantation

**Microstructure (electron-microscopy level): defects in III-V and II-VI semiconductors; defects in crystals; films and epitaxy; interface imperfections; electro-migration; yields; metallization

**Optical: optical properties; displays, solid-state lasers; light-emitting diodes; nonlinear optical materials; optical communications; low loss optical fibers for optical communications; optical modulators; optical storage

**Dielectric: high-voltage dielectrics; high temperature dielectrics; surface effects at semiconductor/insulator interfaces; encapsulation; better capacitors; substrates

Materials

***Elemental and Compound Semiconductors: for electronic circuits; large-scale integration; for displays; for solid-state lasers; for semiconductor memory; variable bandgap; high-temperature semiconductors

***Thin Films: for large-scale integration; for light-emitting diodes; of II-VI compounds; control of metal-lization; thin-film memories; thin-film integrated optical devices; epitaxy; perfection of thin films; bubble memories

**Ceramics: substrates, oxide layers, dielectrics; integrated optics; encapsulation; laser windows

Glasses and Amorphous Materials: optical transmission; integrated optics; laser windows; amorphous semi-conductors; radiation-hard switches; radiation damage; glass for passivation; glass/metal seals

Inorganic, Nonmetallic Elements and Compounds: electro-optic microelectronics; sensors; displays; modulators; detectors; bubble memories

Processes

***Vapor and Electrodeposition, Epitaxy:** yield and process-ing of large-scale integrated circuits; thin film quality; epitaxy; greater miniaturization; control of metallization

***Chemical:** corrosion; compatibility in environment; contacts; connectors; doping; distribution of dopants; etching

Extraction, Purification, Refining: purification; syn-thesis; characterization; high purity optical glasses

Synthesis and Polymerization: encapsulants; conducting adhesives; coatings; seals

Solidification and Crystal Growth: larger, more perfect crystals; monolithic processing of III-V's

Radiation Treatment: ion implantation; radiation damage

Plating and Coating: encapsulation; environmental protection

Consumer Goods

Materials

***Plastics:** stronger plastics; wear resistance; less brittleness; non-flammable plastics; impact-resistant plastics; biodegradable plastics; high-impact foams; conducting plastics; semiconducting plastics

Adhesives, Coatings, Finishes, Seals: Resistant polymers and rubbers; corrosion protection; enamels; hot-water tank coatings; self-cleaning coatings for ranges; reduce permeability of packaging films; bonding; fastening; adhesive joining of fabrics

**Wood and Paper: Wet strength of corrugated paper;
fireproof paper

Processes

***Plastics Extrusion and Moldings: reinforced plastics;
composites; shaping and forming; fabrication ease;
colloid properties; improved fibers

**Synthesis and Polymerization: composite processing;
biodegradable polymers; improved cross-linking;
molecular architecture for special properties; special
visco-elastic properties; improve fiber strength by
controlling molecular orientation

Defense and Space

Properties

***Mechanical and Acoustic: higher strength/weight; light-
weight armor; high strength; mechanical properties of
composites; high-temperature materials; fatigue;
corrosion fatigue; crack propagation; high-temperature
fatigue; creep resistance; fracture toughness; impact
resistance; fatigue resistance; undersea equipment;
materials for pressure hulls

**Microstructure, Electron Microscope Level: dispersion
hardening; microstructural stability; corrosion pitting;
uniformity of mechanical properties; radiation-resistant
materials; hydrogen compatibility

Materials

**Nonferrous Structural Metals and Alloys: improved mechan-
ical properties of structural metals and alloys (see
above)

**Plastics: high-strength plastics; superstrength plastics
for naval vessel superstructures; high-strength fibers

**Composites: composites for ship construction; structural
designs for composites; improved fracture toughness of
composites; fatigue-resistant composites; dispersion-
hardened alloys; reliability of composites

**Adhesives, Coatings, Finishes, Seals: high-temperature coatings; fabrication of metal-nonmetal systems; integrity of polymer adhesives; degradation of adhesive bonds; antifouling coatings for ships; coatings to reduce corrosion; low drag and low contamination paints; room-temperature curing adhesives; thermal-control coatings; ablation materials; cements; sealants for deep-sea equipment

Processes

**Joining: welding of titanium; weldable aluminum alloys; welding of dispersion-hardened alloys; joining of composites; adhesion mechanisms; seals for undersea repeaters

**Testing and Nondestructive Testing: failure analysis; service life; failure prediction; nondestructive testing for welds

Energy

Properties

***Chemical: batteries; higher energy density; improved electrodes; lower weight; longer life; catalysts for batteries; new container materials for batteries; corrosion of cables, of heat exchangers, of turbine blades; radiation effects on corrosion; high-temperature corrosion

**Atomic Structure: solid-state electrolytes; hydrogen embrittlement; super-conducting materials for power transmission

**Microstructure (Electron Microscope Level): Radiation resistance; swelling; void formation; blistering; stability under high neutron fluxes; radiation-hard control equipment

**Thermodynamic: combustion efficiency; thermoelectric power converters; magnetohydrodynamic conversion systems; electrohydrodynamic conversion systems

**Mechanical and Acoustical: high-temperature materials for reactors, both for fuel containers and converters; lightweight conductors; high-temperature alloys for turbines; high-temperature bearings; creep; fracture toughness; high strength; toughness; notch sensitivity; fracture propagation in pipeline materials

Materials

**Ceramics: high-temperature materials for burners; for plasma containment; high-voltage insulators; ceramics for turbine blades

Processes

**Testing and Nondestructive Testing: failure criteria; lifetime prediction; nondestructive testing of reactor components

Environmental Quality

Properties

**Chemical: catalysts for automobile exhausts; pollution detection of control systems; improved, cleaner extraction processes; improved benefication of ores; recovery processes

**Biological: air quality, water quality, land pollution; health hazards; noise; handling corrosive, toxic and dusty materials; biodegradable plastics

Materials

**Plastics: flammability; toxicity; reproducibility of properties; flame retardant; wear; recyclable polymers; scrap polymers used as fuels

Processes

**Extraction, Purification, Refining: improved extraction methods; improved incineration methods; recovery and recycling of wastes; control of pollution and environmental degradation caused by mining and extraction; develop sorting mechanisms and recovery procedures for scrap

Health Services

Properties

***Chemical: understand enzymes, proteins and nuclides; effects of drugs, stimulants and depressants; corrosion of implants; microbial corrosion; stress corrosion

***Biological: biological response to implants; biocompatibility; rejection; toxicity; immunological response

**Microstructure (Electron Microscope Level): adhesion; prosthesis/tissue interface; adhesion between implants and tissue

**Mechanical and Acoustical: artificial bone, teeth, tissue, membranes and organs, better filling material for teeth; fatigue; wear; alloys for joints; high strength

Materials

***Plastics: membranes; artificial teeth; dental adhesives; artificial heart valves; encapsulants for implants; containers for blood

***Prosthetic and Medical Materials: implants; artificial organs, bones, teeth, tissue and membranes; compatibility and biological response

**Fibers and Textiles: membranes, fine wires, organ replacements

**Rubbers: artificial organs, tissue, membranes

**Composites: for implants; bone and tooth replacements; joints; pins; matching strength and stiffness

**Organo- and Organometallic Compounds: prothesis-tissue interface; adhesion between bone and tissue; for implants

Processes

**Synthesis and Polymerization: dental adhesives; compatibility; interface between tissue and prosthesis

**Plastics Extrusion and Molding: precision forming; controlled porosity; artificial organs; heart valves; membranes

**Testing and Nondestructive Testing: quality control; methods to evaluate compatibility; characterization of properties of implants; chemical sensors and monitors

Housing and Other Construction

Properties

**Mechanical and Acoustical: low assembly costs; high stiffness and strength; improved "warmth" of plastics; effects of rolling loads on road surfaces; durability

**Chemical: corrosion; atmospheric degradation; flammability; climate effects on pavement; thermal stability

Materials

**Plastics: develop plastics and easy fabrication methods for use in housing; plastic structures

**Composites: multilayer panels; composites for structural uses; blended ceramics in liquid form; low cost; corrosion-resistant concrete; reinforcement composites to replace steel and concrete

**Adhesives, Coatings, Finishes, Seals: sealants, new joining methods; improved enamel for plumbing

Processes

**Plastics Extrusion and Molding: low-cost polymer fabrication methods; new fabrication methods

**Joining: prefabrication methods; joint materials for bridges

Production Equipment

Materials

**Ferrous Metals and Alloys: harder dies; better cutting tools; better saws; rust resistance

**Non-ferrous Structural Metals and Alloys: cold-forming metals; improved wear and fatigue properties

**Lubricants, Oils, Solvents, Cleansers: tribology-lubricants; wear and abrasion resistance

Processes

**Testing and Nondestructive Testing: quality control; fatigue failures

Transportation Equipment

Properties

***Mechanical and Acoustical: improved strength/weight for auto bodies and engines and for aircraft; fatigue; crack propagation; temperature cycling; impact resistance; wear of tires; energy-absorbing systems; better bearings

**Microstructure (Electron Microscope Level): higher strength/weight; super alloys for engines; corrosion resistance; stress corrosion

**Chemical: corrosion resistance; stress corrosion; corrosion fatigue; high-temperature oxidation; catalytic converters for automotive exhaust

Materials

***Adhesives, Coatings, Finishes, Seals: Adhesives and sealants for aircraft; adhesives for automobile bodies, frames, and repairs; coatings for automobile mufflers; coatings for turbine blades; seals for gas turbines; seals for Wankel engines; refractory coatings

***Lubricants, Oils, Solvents, Cleansers: wear, abrasion resistance

**Ferrous Metals and Alloys: improved strength/weight; improved high-temperature properties; corrosion resistance

**Non-ferrous Structural Metals and Alloys: development of titanium and beryllium alloys; superalloys; high-temperature materials for turbine engines

**Plastics: for automobile bodies; composites; higher strength/weight

**Rubbers: wear and reliability of tires; fabrication processes for tires; castable tires; nondestructive testing for tires

**Composites: develop composites for use in engine and bodies of automobiles and aircraft; joining metals

Processes

**Metal Deformation and Processing: more automation; improved casting; nondestructive testing evaluation; improved fabrication methods; new foundry processes; lower cost

Heat Treatment: improved strength/weight; improved strength; high-temperature properties; warping and cracking during heat treatment

Material Removal: improved shaping methods; lower cost; more efficient methods

Joining: fasteners and bonding systems for aircraft and for automobiles; joining methods for composite materials

213

APPENDIX B

LIAISON REPRESENTATIVES, PANELS,
COMMITTEES, AND CONTRIBUTORS

Liaison Representatives to COSMAT

Emanuel Haynes   National Science Foundation

        Charles E. Falk and Arley T. Bever acted in this capacity
during the first two years of the Survey. Harold W. Paxton
also participated in the liaison activities. Jack T.
Sanderson was the overall COSPUP program officer for NSF
during the Survey. Bodo Bartocha and Israel Warshaw played
an important role in the initiation of the Survey.

Franklin P. Huddle   Library of Congress

        Richard A. Carpenter acted in this capacity during the
first eighteen months of the Survey

Emanuel Horowitz   Interagency Council for Materials

        George C. Deutsch and Earl T. Hayes acted in this capacity
during the first two years of the Survey.

C. Martin Stickley   Advanced Research Projects Agency

        Maurice J. Sinnott acted in this capacity during the first
two years of the Survey. Robert A. Huggins played an
important role in the initiation of the Survey.

Panel I - Materials and Society

Melvin Kranzberg (Chairman)   Georgia Institute of Technology
Cyril S. Smith (Co-Chairman)   Massachusetts Institute of Technology
Harold J. Barnett   Washington University
Walter R. Hibbard, Jr.   Owens-Corning Fiberglas Corporation
Richard W. Roberts   National Bureau of Standards

Panel II - The Nature of Materials Science and Engineering

Richard Claassen (Chairman)	Sandia Laboratories
Joseph D. Andrade	University of Utah
Albert G. H. Dietz	Massachusetts Institute of Technology
Harold E. Goeller	Oak Ridge National Laboratory
Julius J. Harwood	Ford Motor Company
Robert A. Huggins	Stanford University
Morton E. Jones	Texas Instruments, Inc.
Hans H. Landsberg	Resources for the Future, Inc.
Alan M. Lovelace	United States Air Force
Morris Tanenbaum	Western Electric Company, Inc.

Consultant

John P. Howe	Gulf General Atomics, Inc.

Liaison from Data and Information Panel

Lawrence H. Van Vlack	The University of Michigan

Panel III - Institutional Framework of Materials Science and Engineering

Walter R. Hibbard, Jr. (Chairman)	Owens-Corning Fiberglas Corporation
Donald J. Blickwede	Bethlehem Steel Corporation
Paul Chenea	General Motors Corporation
Roger S. Porter	University of Massachusetts
Rustum Roy	Pennsylvania State University
Paul Shewmon	Argonne National Laboratory

Liaison from Data and Information Panel

G. Frederick Bolling	Ford Motor Company

Committee III-A Industry

Paul F. Chenea (Chairman)	General Motors Corporation
Raymond F. Boyer	Dow Chemical Company
Joseph E. Burke	General Electric Company
Jacob E. Goldman	The Xerox Corporation
Fred D. Rosi	Private Consultant
Abe Silverstein	Republic Steel Corporation
Gordon K. Teal	Texas Instruments, Inc.

Consultant

Keith D. Gardels	General Motors Corporation

Committee III-B Government

Paul Shewmon (Chairman)	Argonne National Laboratory
Thomas A. Henrie	Department of the Interior
Robb M. Thomson	National Bureau of Standards

Committee III-C Education

Rustum Roy (Chairman)	Pennsylvania State University
Fred W. Billmeyer	Renssalear Polytechnic Institute
Angel G. Jordan	Carnegie-Mellon University
Robert Maddin	University of Pennsylvania
Walter S. Owen	Northwestern University
J. Robert Schrieffer	University of Pennsylvania

Committee III-D Professional Activities and Manpower

Donald J. Blickwede (Chairman)	Bethlehem Steel Corporation
John D. Alden	Engineering Manpower Commission
Milton Harris	Private Consultant
Roger S. Porter	University of Massachusetts
A. R. Putnam	American Society for Metals
Karl Schwartzwalder	A. C. Spark Plug Division

Panel IV - World Activities in Materials Science and Engineering

W. O. Baker (Chairman)	Bell Telephone Laboratories, Inc.
N. Bruce Hannay	Bell Telephone Laboratories, Inc.
Franklin P. Huddle	Library of Congress
Herman F. Mark	Polytechnic Institute of Brooklyn
Earl R. Parker	University of California, Berkeley
David Swan	Kennecott Copper Corporation
Kyle Wing	International Telephone and Telegraph, Inc.

Consultant

Nathan E. Promisel	National Materials Advisory Board (NAS-NAE-NRC)

Liaison from Data and Information Panel

Robert I. Jaffee	Battelle Memorial Institute

Panel V - Lessons and Opportunities Within Materials Science and Engineering

Daniel C. Drucker	University of Illinois
N. Bruce Hannay	Bell Telephone Laboratories, Inc.

Panel VI - National Objectives and the Contribution of Materials Science and Engineering

Hans H. Landsberg	Resources for the Future, Inc.
Roland W. Schmitt	General Electric Company

Liaison from Data and Information Panel

Kenneth A. Jackson	Bell Telephone Laboratories, Inc.

Data and Information Panel (DIP)

Robert I. Jaffee (Chairman)	Battelle Memorial Institute
G. Frederick Bolling	Ford Motor Company
John C. Fisher	General Electric Company
Kenneth A. Jackson	Bell Telephone Laboratories, Inc.
Robert W. Keyes	IBM, Inc.
Humboldt W. Leverenz	RCA Corporation
Lawrence H. Van Vlack	The University of Michigan

Consultants

James D. Livingston	General Electric Company
William M. Perry	Bethlehem Steel Corporation
Anthony R. Pierce	Bell Telephone Laboratories, Inc.
Myron Wish	Bell Telephone Laboratories, Inc.

Liaison Representatives

James H. Brown	National Science Foundation
George C. Deutsch	National Aeronautics and Space Administration
Franklin P. Huddle	Library of Congress
Leonard L. Lederman	National Science Foundation
Paul Zinner	Department of the Interior

Programmers

Mary Campagnucci
Anne B. Kane

Secretaries

Diane M. Laverty
Catharine E. Little
Judith M. Marshall
Marguerite A. Meyer
Jasmine C. Smith

Individual Contributors

Numerous other persons from both the United States and abroad also contributed to the COSMAT study by providing information, preparing position papers, undertaking data analysis, reviewing reports, and responding to various questionnaires. Their cooperation has proved invaluable. The scope of these contributions will be given in the main Survey Report.

U.S. Technical Societies Addressed for Information

American Ceramic Society
American Chemical Society
American Concrete Institute
American Foundrymen's Society
American Institute of Aeronautics and Astronautics
American Institute of Chemical Engineers
American Iron and Steel Institute
American Nuclear Society
American Physical Society
American Society of Civil Engineers
American Society of Mechanical Engineers
American Society for Metals
American Society for Nondestructive Testing
American Society for Quality Control
American Society for Testing and Materials
American Welding Society
Association of Iron and Steel Engineers
Electrochemical Society
Electron Microscopy Society of America
Federation of Societies for Paint Technology
Forest Products Research Society
Institute of Electrical and Electronic Engineers
Instrument Society of America
National Association of Corrosion Engineers
National Association of Power Engineers
Optical Society of America
Society of Aerospace Material and Process Engineers
Society of Automotive Engineers
Society for Experimental Stress Analysis
Society of Manufacturing Engineers
Society of Plastics Engineers